STUDENT RESOURCE MANUAL

Contemporary Mathematics for Business and Consumers

Robert A. Brechner
Miami-Dade Community College

THE DRYDEN PRESS
HARCOURT BRACE COLLEGE PUBLISHERS
Fort Worth Philadelphia San Diego New York Orlando Austin San Antonio
Toronto Montreal London Sydney Tokyo

Portions of this work were published in previous editions.

Address for orders:
The Dryden Press
6277 Sea Harbor Drive
Orlando, FL 32887-6777
1-800-782-4479

Address for editorial correspondence:
The Dryden Press
301 Commerce Street, Suite 3700
Fort Worth, TX 76102

ISBN: 0-03-018159-3

Printed in the United States of America

6 7 8 9 0 1 2 3 4 5 023 9 8 7 6 5 4 3 2 1

The Dryden Press
Harcourt Brace College Publishers

Contents

Formula Recap Charts

CHAPTER 6 ■ PERCENTS

FORMULAS

Portion = Rate × Base
Rate = Portion ÷ Base
Base = Portion ÷ Rate

CHAPTER 7 ■ INVOICES, TRADE AND CASH DISCOUNTS

FORMULAS

The Invoice

Extended total = Number of items × Cost per item

Trade Discounts—Single

Trade discount amount = List price × Trade discount percent

Net price = List price − Trade discount amount

Net price = List price (100% − Trade discount percent)

$$\text{Trade discount rate} = \frac{\text{Amount of trade discount}}{\text{List Price}}$$

Trade Discounts—Series

Net price = List price × Net price factor

Single equivalent discount = 1 − Net price factor

Trade discount amount = List price × Single equivalent discount

Cash Discounts and Terms of Sale

Net amount due = Net price (100% − Cash discount percent)

$$\text{Partial payment credit} = \frac{\text{Partial payment}}{100\% - \text{Cash discount percent}}$$

Net amount due = Net price − Partial payment credit

CHAPTER 8 ■ MARKUP AND MARKDOWN

FORMULAS

Selling price = Cost + Markup

Cost = Selling price − Markup

Markup = Selling price − Cost

$$\text{Percent markup}_{\text{COST}} = \frac{\text{Markup}}{\text{Cost}}$$

$$\text{Percent markup}_{SP} = \frac{\text{Markup}}{\text{Selling price}}$$

$$\text{Selling price} = \text{Cost}(100\% + \%\text{Markup}_{\text{COST}})$$

$$\text{Cost} = \frac{\text{Selling price}}{100\% + \%\text{Markup}_{\text{COST}}}$$

$$\text{Selling price} = \frac{\text{Cost}}{100\% - \%\text{Markup}_{SP}}$$

$$\text{Cost} = \text{Selling price}(100\% - \%\text{Markup}_{SP})$$

$$\%\text{Markup}_{SP} = \frac{\%\text{Markup}_{\text{COST}}}{100\% + \%\text{Markup}_{\text{COST}}}$$

$$\%\text{Markup}_{\text{COST}} = \frac{\%\text{Markup}_{SP}}{100\% - \%\text{Markup}_{SP}}$$

$$\text{Markdown} = \text{Original price} - \text{Sale price}$$

$$\text{Sale price} = \text{Original price} - \text{Markdown}$$

$$\text{Original price} = \text{Sale price} + \text{Markdown}$$

$$\text{Markdown\%} = \frac{\text{Markdown}}{\text{Original price}}$$

$$\text{Sale price} = \text{Original price}(100\% - \text{Markdown\%})$$

$$\text{Original price} = \frac{\text{Sale price}}{100\% - \text{Markdown\%}}$$

$$\text{Selling price}_{\text{Perishables}} = \frac{\text{Expected selling price}}{\text{Total quantity} - \text{Spoilage}}$$

CHAPTER 9 ■ PAYROLL

FORMULAS

Hourly Wages

Regular pay = Hourly rate × Hours worked

Overtime rate = Hourly rate × Overtime factor

Overtime pay = Overtime rate × Hours worked

Total gross pay = Regular pay + Overtime pay

Piecework

Total gross pay = Output quantity × Rate per unit

Commission

Total gross pay = Total sales × Commission rate

Payroll Deductions

Total deductions = Social security + Medicare + Income Tax + Voluntary deductions

Net pay = Gross pay − Total deductions

Fringe Benefits

$$\text{Fringe benefit percent} = \frac{\text{Total fringe benefits}}{\text{Gross payroll}}$$

Estimated Quarterly Tax

$$\text{Quarterly earnings} = \frac{\text{Estimated annual earnings}}{\text{4 (quarters)}}$$

CHAPTER 10 ■ SIMPLE INTEREST AND PROMISSORY NOTES

FORMULAS

$$\text{Interest} = \text{Principal} \times \text{Rate} \times \text{Time}$$

$$\text{Exact time} = \frac{\text{Number of days of a loan}}{365}$$

$$\text{Ordinary time} = \frac{\text{Number of days of a loan}}{360}$$

$$\text{Maturity value} = \text{Principal} + \text{Interest}$$

$$\text{Maturity value} = \text{Principal}\ (1 + \text{Rate} \times \text{Time})$$

$$\text{Principal} = \frac{\text{Interest}}{\text{Rate} \times \text{Time}}$$

$$\text{Rate} = \frac{\text{Interest}}{\text{Principal} \times \text{Time}}$$

$$\text{Time} = \frac{\text{Interest}}{\text{Principal} \times \text{Rate}}$$

$$\text{Bank discount} = \text{Face value} \times \text{Discount rate} \times \text{Time}$$

$$\text{Proceeds} = \text{Face value} - \text{Bank discount}$$

$$\text{Effective interest rate} = \frac{\text{Bank discount}}{\text{Proceeds} \times \text{Time}}$$

CHAPTER 11 ■ COMPOUND INTEREST & PRESENT VALUE

FORMULAS

$$\text{Compound interest} = \text{Compound amount} - \text{Principal}$$

$$\text{Compounding periods} = \text{Years} \times \text{Periods per year}$$

$$\text{Interest rate per period} = \frac{\text{Nominal rate}}{\text{Periods per year}}$$

$$\text{Compound amount} = \text{Table factor} \times \text{Principal}$$

$$\text{Effective rate} = \frac{\text{Total interest earned in 1st year}}{\text{Principal}}$$

$$\text{Compound amount} = \text{Principal}(1 + \text{interest})^{\text{periods}}$$

$$\text{Present value} = \text{Table factor} \times \text{Compound amount}$$

$$\text{Present value} = \frac{\text{Compound amount}}{(1 + \text{interest})^{\text{periods}}}$$

CHAPTER 12 ■ ANNUITIES

FORMULAS

Future value (ordinary annuity) = Ordinary annuity table factor × Annuity payment

$$FV \text{ (ordinary annuity)} = \text{Payment} \times \frac{(1+i)^n - 1}{i}$$

Future value (annuity due) = Annuity due table factor × Annuity payment

$$FV \text{ (annuity due)} = \text{Payment} \times \frac{(1+i)^n - 1}{i} \times (1+i)$$

Present value (ordinary annuity) = Ordinary annuity table factor × Annuity payment

$$PV \text{ (ordinary annuity)} = \text{Payment} \times \frac{1-(1+i)^{-n}}{i}$$

Present value (annuity due) = Annuity due table factor × Annuity payment

$$PV \text{ (annuity due)} = \text{Payment} \times \frac{1-(1+i)^{-n}}{i} \times (1+i)$$

$$\text{Sinking fund payment} = \frac{\text{Future value of the sinking fund}}{\text{Future value table factor}}$$

$$\text{Sinking fund payment} = \text{Future value} \times \frac{i}{(1+i)^n - 1}$$

$$\text{Amortization payment} = \frac{\text{Original amount of obligation}}{\text{Present value table factor}}$$

$$\text{Amortization payment} = \text{Present value} \times \frac{i}{1-(1+i)^{-n}}$$

CHAPTER 13 ■ CONSUMER AND BUSINESS CREDIT

FORMULAS

$$\text{Periodic rate} = \frac{\text{Annual percentage rate}}{12}$$

Finance charge = Previous month's balance × Periodic rate

$$\text{Average daily balance} = \frac{\text{Sum of daily balances}}{\text{Days in billing cycle}}$$

Finance charge = Average daily balance × Periodic rate

$$\text{New balance} = \text{Previous balance} + \text{Finance charge} + \text{Purchases and cash advances} - \text{Payments and credits}$$

Amount financed = Purchase price − Down payment

Down payment = Purchase price × Down payment percent

Amount financed = Purchase price (100% − Down payment percent)

Total amount of installment payments = Amount financed + Finance charge

Finance charge = Total amount of installment payments − Amount financed

$$\text{Total amount of installment payments} = \text{Monthly payment amount} \times \text{Number of monthly payments}$$

Total deferred payment price = Total of installment payments + Down payment

$$\underset{(\textit{finance charge})}{\text{Interest}} = \underset{(\textit{amount financed})}{\text{Principal}} \times \text{Rate} \times \text{Time}$$

$$\text{Regular monthly payments} = \frac{\text{Total of installment payments}}{\text{Number of months of loan}}$$

$$\text{APR} = \frac{72I}{3P(n+1) + I(n-1)}$$

$$\text{Finance charge} = \frac{\text{Amount financed} \times \text{APR table factor}}{100}$$

$$\text{Sum of digits} = \frac{n(n+1)}{2}$$

$$\text{Rebate fraction} = \frac{\text{Sum of digits of remaining payments}}{\text{Sum of digits of total payment}}$$

Finance charge rebate = Rebate fraction × Total finance charge

Loan payoff = (Payments remaining × Payment amount) – Finance charge rebate

CHAPTER 14 ■ MORTGAGES

FORMULAS

Monthly payment = Number of \$1,000s financed × Table 15-1 factor

Total interest = (Monthly payment × Number of payments) – Amount financed

$$\text{Monthly taxes and insurance (TI)} = \frac{\text{Estimated property tax + Hazard insurance}}{12}$$

Monthly PITI = Monthly PI + Monthly TI

ARM-Calculated interest rate = Index rate + Lender's margin

ARM-Maximum rate per adjustment period = Previous rate + Periodic cap

ARM-Maximum overall rate = Initial rate + Overall cap

Percentage of appraised value = Appraised value × Lender's percentage

Second mortgage potential credit = Percentage of appraised value – First mtg. balance

$$\text{Housing expense ratio} = \frac{\text{Monthly housing expense (PITI)}}{\text{Monthly gross income}}$$

$$\text{Total obligations ratio} = \frac{\text{Total monthly financial obligations}}{\text{Monthly gross income}}$$

CHAPTER 15 ■ FINANCIAL STATEMENTS AND RATIOS

FORMULAS

Liquidity Ratios

Working capital = Current assets – Current liabilities

$$\text{Current ratio} = \frac{\text{Current assets}}{\text{Current liabilities}}$$

Quick assets = Cash + Marketable securities + Receivables

$$\text{Acid test ratio} = \frac{\text{Quick assets}}{\text{Current liabilities}}$$

Efficiency Ratios

$$\text{Average collection period} = \frac{\text{Accounts receivable}}{\text{Credit sales/365}}$$

$$\text{Average inventory} = \frac{\text{Beginning inventory} + \text{Ending inventory}}{2}$$

$$\text{Inventory turnover} = \frac{\text{Cost of goods sold}}{\text{Average inventory}}$$

$$\text{Asset turnover ratio} = \frac{\text{Net sales}}{\text{Total assets}}$$

Leverage Ratios

$$\text{Debt-to-assets ratio} = \frac{\text{Total liabilities}}{\text{Total assets}}$$

$$\text{Debt-to-equity ratio} = \frac{\text{Total liabilities}}{\text{Owner's equity}}$$

Profitability Ratios

$$\text{Gross profit margin} = \frac{\text{Gross profit}}{\text{Net sales}}$$

$$\text{Net profit margin} = \frac{\text{Net income}}{\text{Net sales}}$$

$$\text{Return on investment} = \frac{\text{Net income}}{\text{Owner's equity}}$$

CHAPTER 16 ■ INVENTORY

FORMULAS

Inventory Valuation—Average Cost Method

$$\text{Average cost per unit} = \frac{\text{Cost of goods available for sale}}{\text{Total units available for sale}}$$

Ending inventory = Units in ending inventory × Average cost per unit

Inventory Estimation—Retail Method

$$\text{Cost ratio} = \frac{\text{Goods available for sale at cost}}{\text{Goods available for sale at retail}}$$

Estimated ending inventory at cost = Ending inventory at retail × Cost ratio

Inventory Estimation—Gross Profit Method

Estimated cost of goods sold = Net sales (100% − Gross margin %)

Inventory Turnover—Retail

$$\text{Average inventory}_{\text{retail}} = \frac{\text{Beginning inventory}_{\text{retail}} + \text{Ending inventory}_{\text{retail}}}{2}$$

$$\text{Inventory turnover}_{\text{retail}} = \frac{\text{Net sales}}{\text{Average inventory}_{\text{retail}}}$$

Inventory Turnover—Cost

$$\text{Average inventory}_{\text{cost}} = \frac{\text{Beginning inventory}_{\text{cost}} + \text{Ending inventory}_{\text{cost}}}{2}$$

$$\text{Inventory turnover}_{\text{cost}} = \frac{\text{Cost of goods sold}}{\text{Average inventory}_{\text{cost}}}$$

Target Inventory

$$\text{Target average inventory}_{\text{cost}} = \frac{\text{Cost of goods sold}}{\text{Published inventory turnover}_{\text{cost}}}$$

$$\text{Target average inventory}_{\text{retail}} = \frac{\text{Net sales}}{\text{Published inventory turnover}_{\text{retail}}}$$

CHAPTER 17 ■ DEPRECIATION

FORMULAS

Straight-Line Method

Total cost = Cost + Freight + Setup expenses

Total depreciation = Total cost − Salvage value

$$\text{Annual depreciation} = \frac{\text{Total depreciation}}{\text{Estimated useful life (years)}}$$

Sum-of-the-Years' Digits Method

$$\text{SYD depreciation rate fraction} = \frac{\text{Years of useful life remaining}}{\frac{n(n+1)}{2}}$$

Annual depreciation = Total depreciation × Depreciation rate fraction

Declining-Balance Method

$$\text{Declining-balance rate} = \frac{1}{\text{Useful life}} \times \text{Multiple}$$

Beginning book value = Ending book value of the previous year

Ending book value = Beginning book value − Depreciation for the year

Units-of-Production Method

$$\text{Depreciation per unit} = \frac{\text{Cost} - \text{Salvage value}}{\text{Units of useful life}}$$

Annual depreciation = Depreciation per unit × Units produced

MACRS Depreciation

Depreciation deduction = Original basis × Cost recovery percentage

Natural Resource Depletion

$$\text{Average depletion cost per unit} = \frac{\text{Total cost of resource} - \text{Residual value}}{\text{Estimated total units available}}$$

$$\text{Periodic depletion cost} = \text{Units produced in current period} \times \text{Average depletion cost per unit}$$

CHAPTER 18 ■ TAXES

FORMULAS

Sales and Excise Taxes

Sales tax = Selling price × Sales tax rate

Total purchase price = Selling price + Sales tax + Other charges

$$\text{Selling price} = \frac{\text{Total purchase price}}{100\% + \text{Sales tax rate}}$$

Sales tax = Total purchase price − Selling price

Excise tax = Selling price × Excise tax rate

Excise tax = Number of units × Excise tax per unit

Total purchase price = Selling price + Sales tax + Excise tax

Property Tax

a. Expressed as a Percent

Property tax = Assessed value of property × Tax rate

b. Expressed per $100 of Assessed Value

Property tax = Number of $100 of assessed value × Tax per $100

c. Expressed per $1,000 of Assessed Value

Property tax = Number of $1,000 of assessed value × Tax per $1,000

d. Expressed in Mills

Tax rate in decimal form = Tax rate in mills × .001

Property tax = Assessed value × Tax rate in decimal form

Community Tax Rate

$$\text{Tax rate per dollar (decimal form)} = \frac{\text{Total taxes required}}{\text{Total assessed property value}}$$

Income Tax

Refund (−) or amount owed (+) = Tax liability − Credits + Other taxes − Payments

CHAPTER 19 ■ INSURANCE

FORMULAS

Life Insurance

$$\text{Number of \$1,000} = \frac{\text{Face value of policy}}{1{,}000}$$

Annual premium = Number of $1,000 × Rate per $1,000

Premium other than annual = Annual premium × Premium factor

Income shortfall = Total living expenses − Total income

$$\text{Insurance needed} = \frac{\text{Income shortfall}}{\text{Prevailing interest rate}}$$

Property Insurance

Total annual fire premium = Building premium + Contents premium

Short-rate premium = Annual premium × Short-rate

Short-rate refund = Annual premium − Short-rate premium

Regular refund = Annual premium − Premium for period in force

$$\text{Coinsurance coverage ratio} = \frac{\text{Insurance carried}}{\text{Insurance required}}$$

$$\text{Amount of loss paid by insurance} = \frac{\text{Insurance carried}}{\text{Insurance required}} \times \text{Amount of loss}$$

$$\text{Carrier's percent of total coverage} = \frac{\text{Amount of carrier's policy}}{\text{Total amount of insurance}}$$

Carrier's share of loss = Amount of loss × Carrier's percent of total coverage

CHAPTER 20 ■ INVESTMENTS

FORMULAS

Stocks

Dividend per share (preferred) = Par value × Dividend rate

$$\text{Dividend per share (common)} = \frac{\text{Total common dividend}}{\text{Number of shares (common)}}$$

$$\text{Current yield} = \frac{\text{Annual dividend per share}}{\text{Current price of the stock}}$$

$$\text{Price-earnings ratio} = \frac{\text{Current price per share}}{\text{Earnings per share}}$$

Gain (or loss) on stock = Proceeds − Total cost

Bonds

Price per bond = Current market price + Accrued interest + Commission

Proceeds = Current market price + Accrued interest − Commission

$$\text{Current yield} = \frac{\text{Annual interest}}{\text{Current market price}}$$

Mutual funds

Mutual fund sales charge = Offer price − Net asset value

$$\text{Sales charge percent} = \frac{\text{Sales charge}}{\text{Net asset value}}$$

$$\text{Net asset value (NAV)} = \frac{\text{Total assets} - \text{Total liabilities}}{\text{Number of shares outstanding}}$$

$$\text{Number of shares purchased} = \frac{\text{Total investment}}{\text{Offer price}}$$

$$\text{Return on investment (ROI)} = \frac{\text{Total gain}}{\text{Total cost of purchase}}$$

FORMULAS

Ungrouped Data

$$\text{Mean} = \frac{\text{Sum of values}}{\text{Number of values}}$$

Median (odd number of values) = Middle value

$$\text{Median (even number of values)} = \frac{\text{Middle value} + \text{Middle value}}{2}$$

Mode = Value or values that occur most frequently

Range = Highest value − Lowest value

Grouped Data

$$\text{Mean} = \frac{\text{Sum of (frequency} \times \text{midpoint)}}{\text{Sum of frequency}}$$

Charts, Tables, and Exhibits

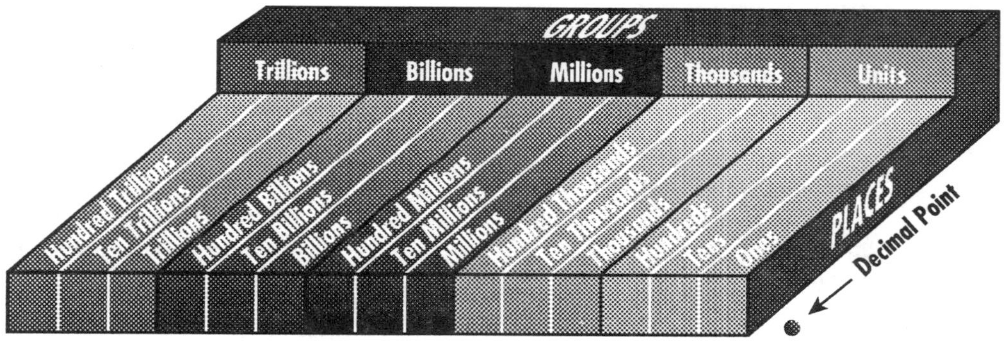

EXHIBIT 1-1

WHOLE NUMBER PLACE VALUE CHART

Rules of Divisibility

A Number Is Divisible By	Conditions
2	If the last digit is 0, 2, 4, 6, or 8.
3	If the sum of the digits is divisible by 3.
4	If the last two digits are divisible by 4.
5	If the last digit is 0 or 5.
6	If the number is divisible by 2 and 3, or if it is even and the sum of the digits is divisible by 3.
8	If the last three digits are divisible by 8.
9	If the sum of the digits is divisible by 9.
10	If the last digit is 0.

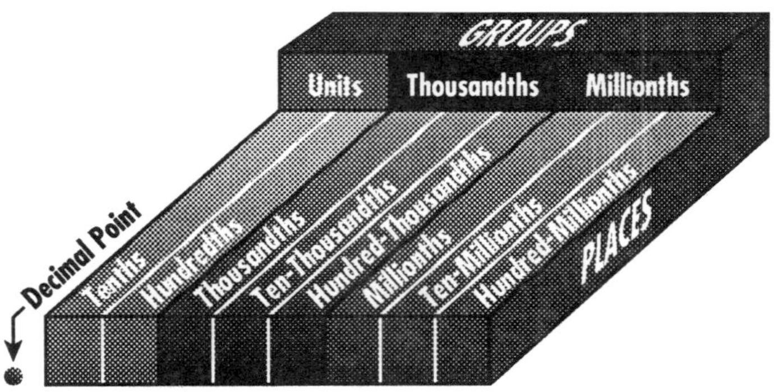

EXHIBIT 3-1

DECIMAL NUMBERS PLACE VALUE CHART

CHECKBOOK BALANCE	$	STATEMENT BALANCE	$
Deduct: Service Charges & Other Debits		Deduct: Outstanding Checks	
SUB TOTAL		SUB TOTAL	
Add: Interest Earned & Other Credits		Add: Deposits in Transit	
ADJUSTED CHECKBOOK BALANCE		ADJUSTED STATEMENT BALANCE	

Checks Outstanding		
No.	Amount	
Total		

EXHIBIT 4-14

BANK STATEMENT RECONCILIATION FORM

Key Words and Phrases for Creating Equations

Equal Sign:
is, are, was, equals, gives, giving, leaves, leaving, makes, denotes

Addition:
added to, totals, the sum of, plus, more than, increased by, greater than, and

Subtraction:
less, less than, minus, difference between, decreased by, reduced by

Multiplication:
of, times, product of, multiplied by, twice, double, triple, at, @

Division:
divide, divided by, the average of, divided into, the quotient of

The Magic Triangle

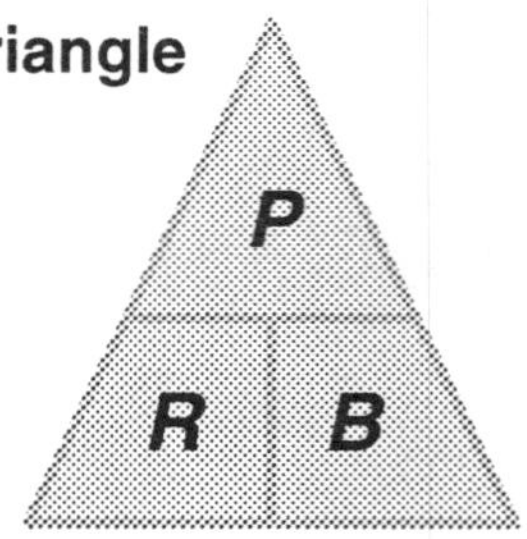

The triangle is divided into three sections, representing the portion, rate, and base. By circling or covering the letter in the triangle that corresponds to the unknown of the problem, the triangle will "magically" reveal the correct formula to use.

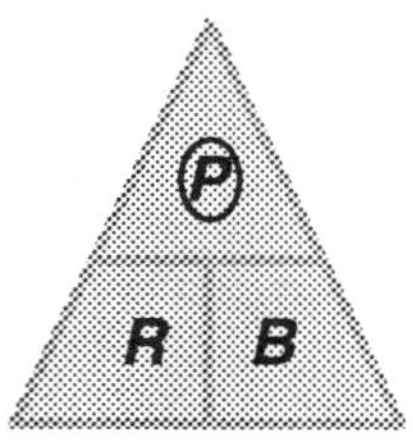

$P = R \times B$

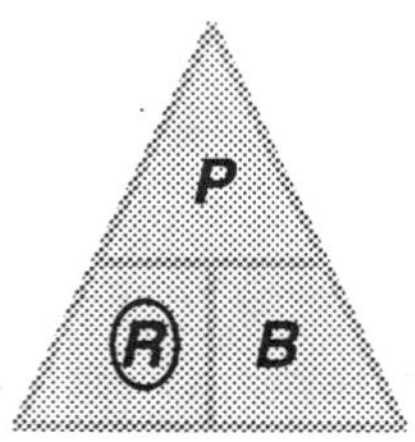

$R = \frac{P}{B}$

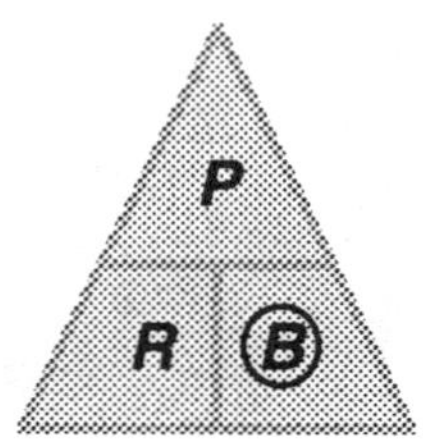

$B = \frac{P}{R}$

EXHIBIT 7-1

TYPICAL INVOICE FORMAT

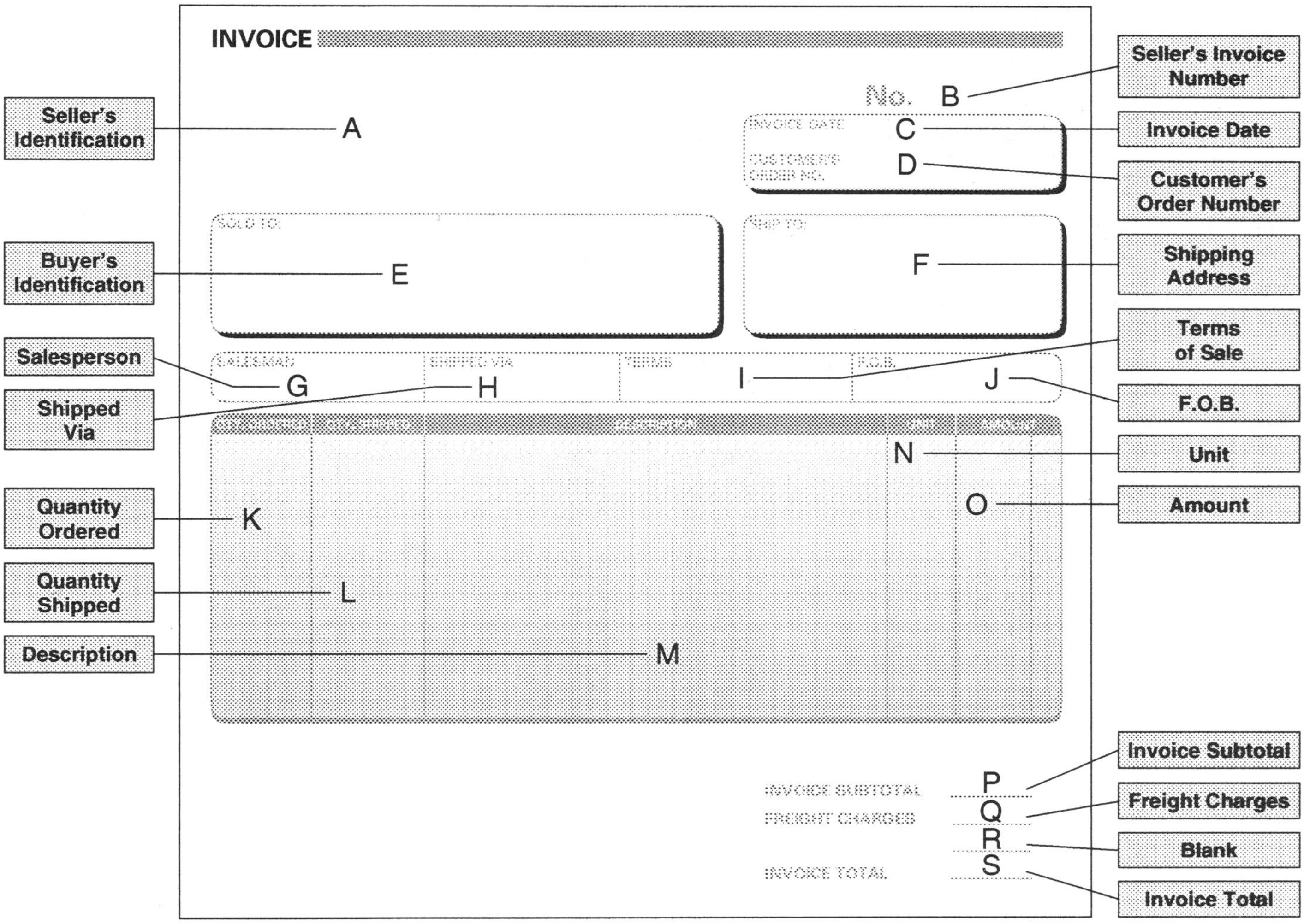

INVOICE TERMINOLOGY

A **Seller's Identification**—Name, address, and logo or corporate symbol of the seller

B **Seller's Invoice Number**—Seller's identification number of the transaction

C **Invoice Date**—Date the invoice was written

D **Customer's Order Number**—Buyer's identification number of the transaction

E **Buyer's Identification**—Name and mailing address of the buyer

F **Shipping Address**—Address where merchandise will be shipped

G **Salesperson**—Name of salesperson credited with the sale

H **Shipped Via**—Name of freight company handling the shipment

I **Terms**—Terms of Sale—Section detailing date of payment and cash discount

J **F.O.B.**—"Free on Board"—Section detailing who pays the freight charges

K **Quantity Ordered**—Number of units ordered

L **Quantity Shipped**—Number of units shipped

M **Description**—Detailed description of the merchandise, including model numbers

N **Unit**—Price per unit of merchandise

O **Amount**—Extended total—Quantity in units times the unit price for each line

P **Invoice Subtotal**—Total of the "amount" column—Merchandise total

Q **Freight Charges**—Shipping charges—Cost to physically transport the merchandise from the seller to the buyer

R **Blank Line**—Line used for other charges, such as insurance or handling

S **Invoice Total**—Total amount of the invoice—Includes merchandise plus all other charges

INVOICE ABBREVIATIONS

ea.	each	pr.	pair	in.	inch	oz	ounce
dz. or doz.	dozen	dm. or drm.	drum	ft	foot	g or gr	gram
gr. or gro.	gross	bbl.	barre	yd	yard	kg	kilogram
bx.	box	sk.	sack	mm	millimeter	pt	pint
cs.	case	@	at	cm	centimeter	qt	quart
ct. or crt.	crate	C.	100 items	m	meter	gal	gallon
ctn. or cart.	carton	M.	1,000 items	lb	pound	cwt	hundred weight

Freight Terminology

EXHIBIT 7-3

FREIGHT TERMINOLOGY

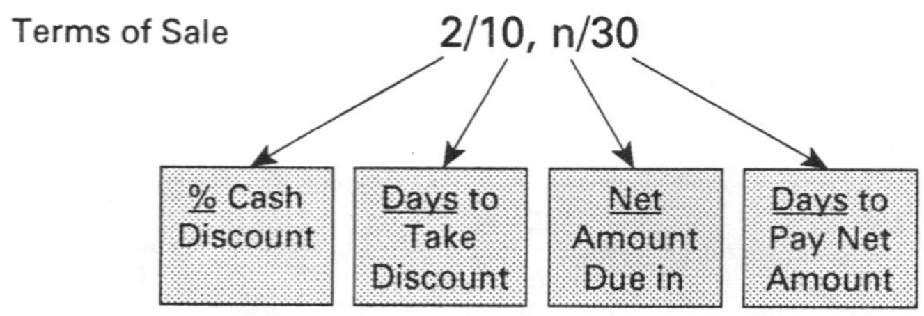

The following illustration shows a time line of the discount period and credit period on an invoice, dated October 15. The 2/10, n/30 terms of sale stipulate a cash discount if the bill is paid within 10 days. If not, the balance is due in 30 days. As you can see, the cash discount period runs for 10 days from the invoice date, October 15 to October 25. The credit period, 30 days, extends from the invoice date through November 14.

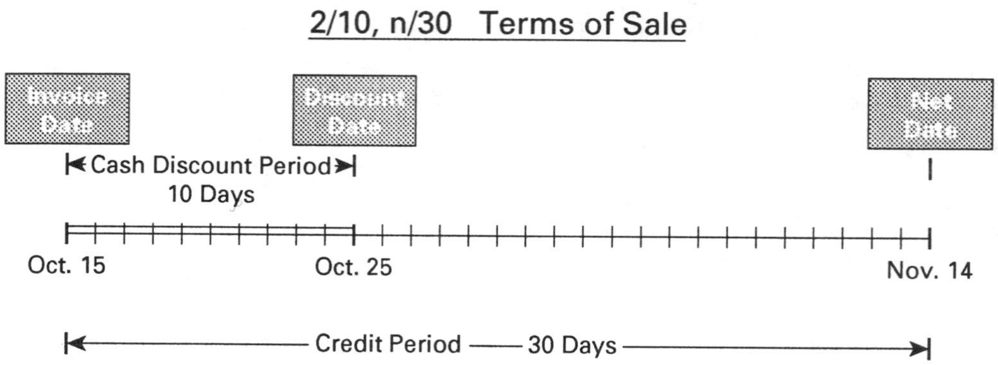

MARRIED Persons—**WEEKLY** Payroll Period

(For Wages Paid in 1996)

If the wages are–		And the number of withholding allowances claimed is—										
At least	But less than	0	1	2	3	4	5	6	7	8	9	10
		The amount of income tax to be withheld is—										
$740	$750	93	86	79	71	64	56	49	42	34	27	20
750	760	95	87	80	73	65	58	51	43	36	29	21
760	770	96	89	82	74	67	59	52	45	37	30	23
770	780	98	90	83	76	68	61	54	46	39	32	24
780	790	99	92	85	77	70	62	55	48	40	33	26
790	800	101	93	86	79	71	64	57	49	42	35	27
800	810	102	95	88	80	73	65	58	51	43	36	29
810	820	104	96	89	82	74	67	60	52	45	38	30
820	830	105	98	91	83	76	68	61	54	46	39	32
830	840	107	99	92	85	77	70	63	55	48	41	33
840	850	108	101	94	86	79	71	64	57	49	42	35
850	860	110	102	95	88	80	73	66	58	51	44	36
860	870	113	104	97	89	82	74	67	60	52	45	38
870	880	116	105	98	91	83	76	69	61	54	47	39
880	890	119	107	100	92	85	77	70	63	55	48	41
890	900	121	108	101	94	86	79	72	64	57	50	42
900	910	124	111	103	95	88	80	73	66	58	51	44
910	920	127	113	104	97	89	82	75	67	60	53	45
920	930	130	116	106	98	91	83	76	69	61	54	47
930	940	133	119	107	100	92	85	78	70	63	56	48
940	950	135	122	109	101	94	86	79	72	64	57	50
950	960	138	125	111	103	95	88	81	73	66	59	51
960	970	141	127	114	104	97	89	82	75	67	60	53
970	980	144	130	116	106	98	91	84	76	69	62	54
980	990	147	133	119	107	100	92	85	78	70	63	56
990	1,000	149	136	122	109	101	94	87	79	72	65	57
1,000	1,010	152	139	125	111	103	95	88	81	73	66	59
1,010	1,020	155	141	128	114	104	97	90	82	75	68	60
1,020	1,030	158	144	130	117	106	98	91	84	76	69	62
1,030	1,040	161	147	133	119	107	100	93	85	78	71	63
1,040	1,050	163	150	136	122	109	101	94	87	79	72	65
1,050	1,060	166	153	139	125	111	103	96	88	81	74	66
1,060	1,070	169	155	142	128	114	104	97	90	82	75	68
1,070	1,080	172	158	144	131	117	106	99	91	84	77	69
1,080	1,090	175	161	147	133	120	107	100	93	85	78	71
1,090	1,100	177	164	150	136	123	109	102	94	87	80	72
1,100	1,110	180	167	153	139	125	112	103	96	88	81	74
1,110	1,120	183	169	156	142	128	114	105	97	90	83	75
1,120	1,130	186	172	158	145	131	117	106	99	91	84	77
1,130	1,140	189	175	161	147	134	120	108	100	93	86	78
1,140	1,150	191	178	164	150	137	123	109	102	94	87	80
1,150	1,160	194	181	167	153	139	126	112	103	96	89	81
1,160	1,170	197	183	170	156	142	128	115	105	97	90	83
1,170	1,180	200	186	172	159	145	131	117	106	99	92	84
1,180	1,190	203	189	175	161	148	134	120	108	100	93	86
1,190	1,200	205	192	178	164	151	137	123	109	102	95	87
1,200	1,210	208	195	181	167	153	140	126	112	103	96	89
1,210	1,220	211	197	184	170	156	142	129	115	105	98	90
1,220	1,230	214	200	186	173	159	145	131	118	106	99	92
1,230	1,240	217	203	189	175	162	148	134	121	108	101	93
1,240	1,250	219	206	192	178	165	151	137	123	110	102	95
1,250	1,260	222	209	195	181	167	154	140	126	112	104	96
1,260	1,270	225	211	198	184	170	156	143	129	115	105	98
1,270	1,280	228	214	200	187	173	159	145	132	118	107	99
1,280	1,290	231	217	203	189	176	162	148	135	121	108	101
1,290	1,300	233	220	206	192	179	165	151	137	124	110	102
1,300	1,310	236	223	209	195	181	168	154	140	126	113	104
1,310	1,320	239	225	212	198	184	170	157	143	129	115	105
1,320	1,330	242	228	214	201	187	173	159	146	132	118	107
1,330	1,340	245	231	217	203	190	176	162	149	135	121	108
1,340	1,350	247	234	220	206	193	179	165	151	138	124	110
1,350	1,360	250	237	223	209	195	182	168	154	140	127	113
1,360	1,370	253	239	226	212	198	184	171	157	143	129	116
1,370	1,380	256	242	228	215	201	187	173	160	146	132	119
1,380	1,390	259	245	231	217	204	190	176	163	149	135	121

$1,390 and over Use Table 1(b) for a **MARRIED person** on page 34. Also see the instructions on page 32.

EXHIBIT 9-2

PAYROLL DEDUCTIONS—MARRIED, PAID WEEKLY

SINGLE Persons—**MONTHLY** Payroll Period

(For Wages Paid in 1996)

If the wages are–		And the number of withholding allowances claimed is—										
At least	But less than	0	1	2	3	4	5	6	7	8	9	10
		The amount of income tax to be withheld is—										
$2,440	$2,480	380	321	272	241	209	177	145	113	81	49	17
2,480	2,520	391	332	278	247	215	183	151	119	87	55	23
2,520	2,560	403	343	284	253	221	189	157	125	93	61	29
2,560	2,600	414	354	295	259	227	195	163	131	99	67	35
2,600	2,640	425	366	306	265	233	201	169	137	105	73	41
2,640	2,680	436	377	317	271	239	207	175	143	111	79	47
2,680	2,720	447	388	328	277	245	213	181	149	117	85	53
2,720	2,760	459	399	340	283	251	219	187	155	123	91	59
2,760	2,800	470	410	351	291	257	225	193	161	129	97	65
2,800	2,840	481	422	362	303	263	231	199	167	135	103	71
2,840	2,880	492	433	373	314	269	237	205	173	141	109	77
2,880	2,920	503	444	384	325	275	243	211	179	147	115	83
2,920	2,960	515	455	396	336	281	249	217	185	153	121	89
2,960	3,000	526	466	407	347	288	255	223	191	159	127	95
3,000	3,040	537	478	418	359	299	261	229	197	165	133	101
3,040	3,080	548	489	429	370	310	267	235	203	171	139	107
3,080	3,120	559	500	440	381	321	273	241	209	177	145	113
3,120	3,160	571	511	452	392	333	279	247	215	183	151	119
3,160	3,200	582	522	463	403	344	285	253	221	189	157	125
3,200	3,240	593	534	474	415	355	296	259	227	195	163	131
3,240	3,280	604	545	485	426	366	307	265	233	201	169	137
3,280	3,320	615	556	496	437	377	318	271	239	207	175	143
3,320	3,360	627	567	508	448	389	329	277	245	213	181	149
3,360	3,400	638	578	519	459	400	340	283	251	219	187	155
3,400	3,440	649	590	530	471	411	352	292	257	225	193	161
3,440	3,480	660	601	541	482	422	363	303	263	231	199	167
3,480	3,520	671	612	552	493	433	374	314	269	237	205	173
3,520	3,560	683	623	564	504	445	385	326	275	243	211	179
3,560	3,600	694	634	575	515	456	396	337	281	249	217	185
3,600	3,640	705	646	586	527	467	408	348	289	255	223	191
3,640	3,680	716	657	597	538	478	419	359	300	261	229	197
3,680	3,720	727	668	608	549	489	430	370	311	267	235	203
3,720	3,760	739	679	620	560	501	441	382	322	273	241	209
3,760	3,800	750	690	631	571	512	452	393	333	279	247	215
3,800	3,840	761	702	642	583	523	464	404	345	285	253	221
3,840	3,880	772	713	653	594	534	475	415	356	296	259	227
3,880	3,920	783	724	664	605	545	486	426	367	307	265	233
3,920	3,960	795	735	676	616	557	497	438	378	319	271	239
3,960	4,000	806	746	687	627	568	508	449	389	330	277	245
4,000	4,040	817	758	698	639	579	520	460	401	341	283	251
4,040	4,080	828	769	709	650	590	531	471	412	352	293	257
4,080	4,120	839	780	720	661	601	542	482	423	363	304	263
4,120	4,160	851	791	732	672	613	553	494	434	375	315	269
4,160	4,200	862	802	743	683	624	564	505	445	386	326	275
4,200	4,240	873	814	754	695	635	576	516	457	397	338	281
4,240	4,280	884	825	765	706	646	587	527	468	408	349	289
4,280	4,320	895	836	776	717	657	598	538	479	419	360	300
4,320	4,360	907	847	788	728	669	609	550	490	431	371	312
4,360	4,400	918	858	799	739	680	620	561	501	442	382	323
4,400	4,440	929	870	810	751	691	632	572	513	453	394	334
4,440	4,480	940	881	821	762	702	643	583	524	464	405	345
4,480	4,520	952	892	832	773	713	654	594	535	475	416	356
4,520	4,560	965	903	844	784	725	665	606	546	487	427	368
4,560	4,600	977	914	855	795	736	676	617	557	498	438	379
4,600	4,640	989	926	866	807	747	688	628	569	509	450	390
4,640	4,680	1,002	937	877	818	758	699	639	580	520	461	401
4,680	4,720	1,014	948	888	829	769	710	650	591	531	472	412
4,720	4,760	1,027	961	900	840	781	721	662	602	543	483	424
4,760	4,800	1,039	973	911	851	792	732	673	613	554	494	435
4,800	4,840	1,051	985	922	863	803	744	684	625	565	506	446
4,840	4,880	1,064	998	933	874	814	755	695	636	576	517	457
4,880	4,920	1,076	1,010	944	885	825	766	706	647	587	528	468
4,920	4,960	1,089	1,023	957	896	837	777	718	658	599	539	480
4,960	5,000	1,101	1,035	969	907	848	788	729	669	610	550	491
5,000	5,040	1,113	1,047	982	919	859	800	740	681	621	562	502
$5,040 and over		Use Table 4(a) for a **SINGLE person** on page 34. Also see the instructions on page 32.										

EXHIBIT 9-3

PAYROLL DEDUCTIONS—
SINGLE, PAID MONTHLY

Payroll Period	One Withholding Allowance
Weekly	$ 49.04
Biweekly	98.08
Semimonthly	106.25
Monthly...............................	212.50
Quarterly	637.50
Semiannually........................	1,275.00
Annually..............................	2,550.00
Daily or miscellaneous (each day of the payroll period)	9.81

EXHIBIT 9-4

PERCENTAGE METHOD WITHHOLDING ALLOWANCES

	Tax Rate	Wage Base
Social Security (OASDI)	6.2%	$62,700
Medicare	1.45%	no limit

	Tax Rate	Wage Base
Social Security	12.4% (6.2% × 2)	$62,700
Medicare	2.9% (1.45% × 2)	No limit

Tables for Percentage Method of Withholding

(For Wages Paid in 1996)

TABLE 1—WEEKLY Payroll Period

(a) SINGLE person (including head of household)—

If the amount of wages (after subtracting withholding allowances) is: Not over $50 . . . . . The amount of income tax to withhold is: $0

Over—	But not over—		of excess over—
$50	—$489	15%	—$50
$489	—$1,033	$65.85 plus 28%	—$489
$1,033	—$2,361	$218.17 plus 31%	—$1,033
$2,361	—$5,100	$629.85 plus 36%	—$2,361
$5,100		$1,615.89 plus 39.6%	—$5,100

(b) MARRIED person—

If the amount of wages (after subtracting withholding allowances) is: Not over $124 . . . . The amount of income tax to withhold is: $0

Over—	But not over—		of excess over—
$124	—$851	15%	—$124
$851	—$1,725	$109.05 plus 28%	—$851
$1,725	—$2,920	$353.77 plus 31%	—$1,725
$2,920	—$5,152	$724.22 plus 36%	—$2,920
$5,152		$1,527.74 plus 39.6%	—$5,152

TABLE 2—BIWEEKLY Payroll Period

(a) SINGLE person (including head of household)—

If the amount of wages (after subtracting withholding allowances) is: Not over $101 . . . . The amount of income tax to withhold is: $0

Over—	But not over—		of excess over—
$101	—$979	15%	—$101
$979	—$2,066	$131.70 plus 28%	—$979
$2,066	—$4,721	$436.06 plus 31%	—$2,066
$4,721	—$10,200	$1,259.11 plus 36%	—$4,721
$10,200		$3,231.55 plus 39.6%	—$10,200

(b) MARRIED person—

If the amount of wages (after subtracting withholding allowances) is: Not over $247 . . . . The amount of income tax to withhold is: $0

Over—	But not over—		of excess over—
$247	—$1,702	15%	—$247
$1,702	—$3,449	$218.25 plus 28%	—$1,702
$3,449	—$5,840	$707.41 plus 31%	—$3,449
$5,840	—$10,304	$1,448.62 plus 36%	—$5,840
$10,304		$3,055.66 plus 39.6%	—$10,304

TABLE 3—SEMIMONTHLY Payroll Period

(a) SINGLE person (including head of household)—

If the amount of wages (after subtracting withholding allowances) is: Not over $109 . . . . The amount of income tax to withhold is: $0

Over—	But not over—		of excess over—
$109	—$1,060	15%	—$109
$1,060	—$2,239	$142.65 plus 28%	—$1,060
$2,239	—$5,115	$472.77 plus 31%	—$2,239
$5,115	—$11,050	$1,364.33 plus 36%	—$5,115
$11,050		$3,500.93 plus 39.6%	—$11,050

(b) MARRIED person—

If the amount of wages (after subtracting withholding allowances) is: Not over $268 . . . . The amount of income tax to withhold is: $0

Over—	But not over—		of excess over—
$268	—$1,844	15%	—$268
$1,844	—$3,736	$236.40 plus 28%	—$1,844
$3,736	—$6,327	$766.16 plus 31%	—$3,736
$6,327	—$11,163	$1,569.37 plus 36%	—$6,327
$11,163		$3,310.33 plus 39.6%	—$11,163

TABLE 4—MONTHLY Payroll Period

(a) SINGLE person (including head of household)—

If the amount of wages (after subtracting withholding allowances) is: Not over $219 . . . . The amount of income tax to withhold is: $0

Over—	But not over—		of excess over—
$219	—$2,121	15%	—$219
$2,121	—$4,477	$285.30 plus 28%	—$2,121
$4,477	—$10,229	$944.98 plus 31%	—$4,477
$10,229	—$22,100	$2,728.10 plus 36%	—$10,229
$22,100		$7,001.66 plus 39.6%	—$22,100

(b) MARRIED person—

If the amount of wages (after subtracting withholding allowances) is: Not over $535 . . . . The amount of income tax to withhold is: $0

Over—	But not over—		of excess over—
$535	—$3,688	15%	—$535
$3,688	—$7,473	$472.95 plus 28%	—$3,688
$7,473	—$12,654	$1,532.75 plus 31%	—$7,473
$12,654	—$22,325	$3,138.86 plus 36%	—$12,654
$22,325		$6,620.42 plus 39.6%	—$22,325

EXHIBIT 9-5

TABLES FOR PERCENTAGE METHOD OF WITHHOLDING

Magic Triangle
Simple Interest Formula

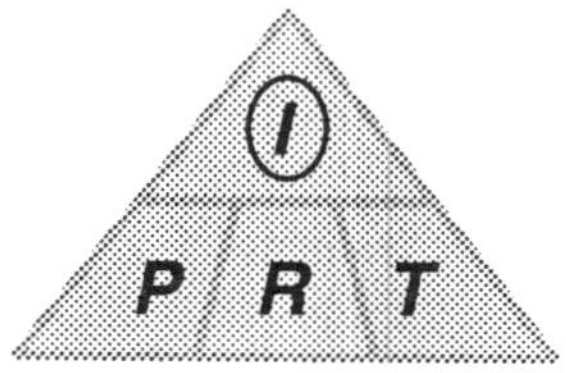

I = PRT

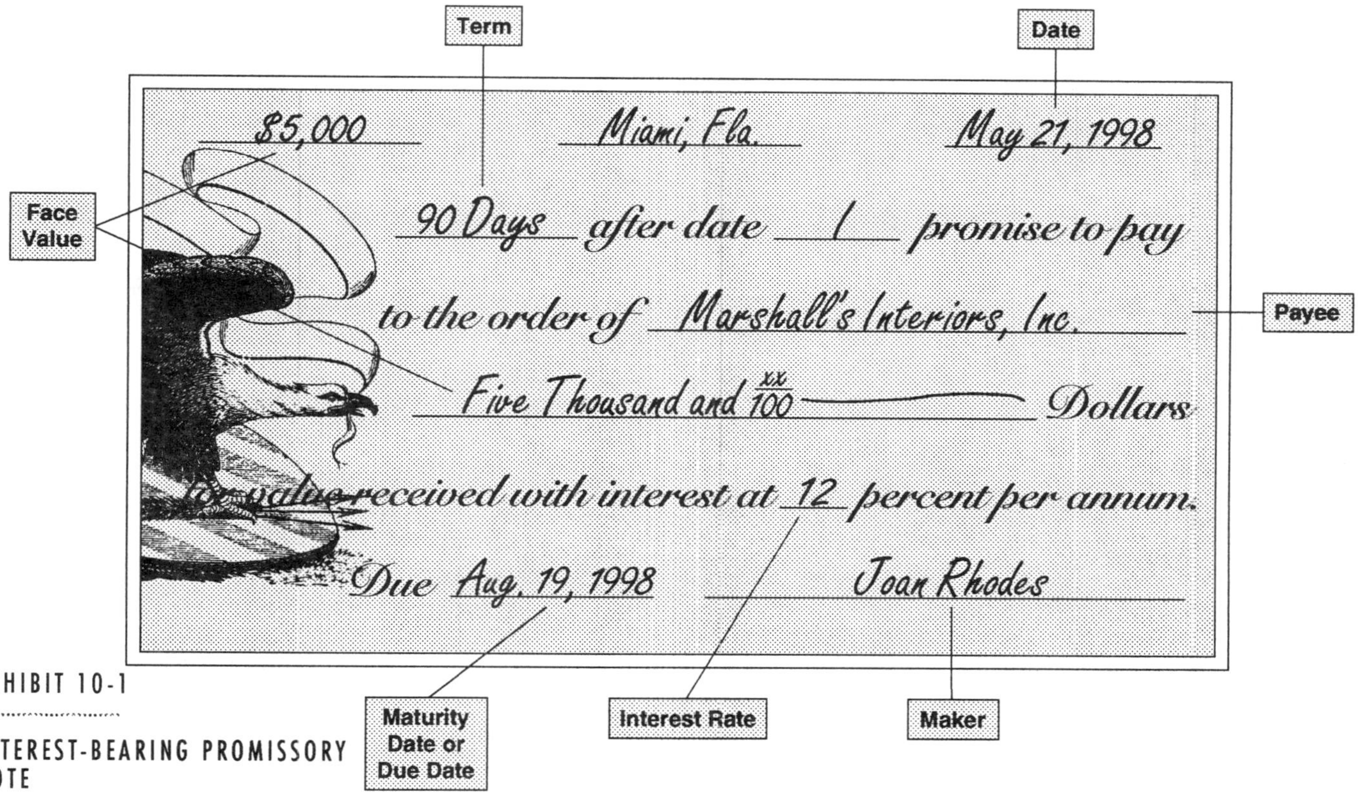

EXHIBIT 10-1

INTEREST-BEARING PROMISSORY NOTE

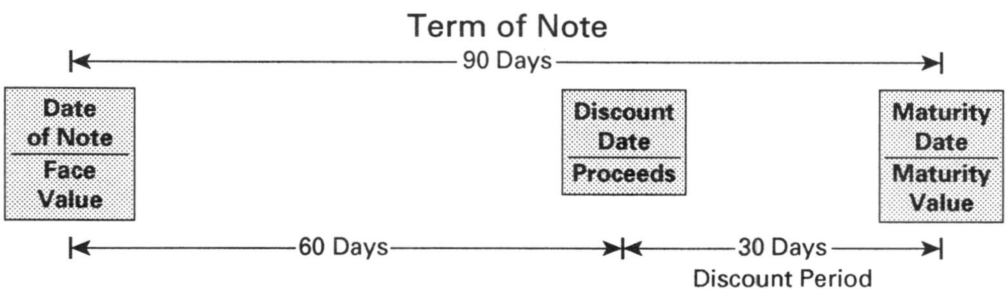

EXHIBIT 10-2

TIME-LINE FOR DISCOUNTED NOTE

EXHIBIT 11-2

PRESENT VALUE AND FUTURE VALUE AT COMPOUND INTEREST

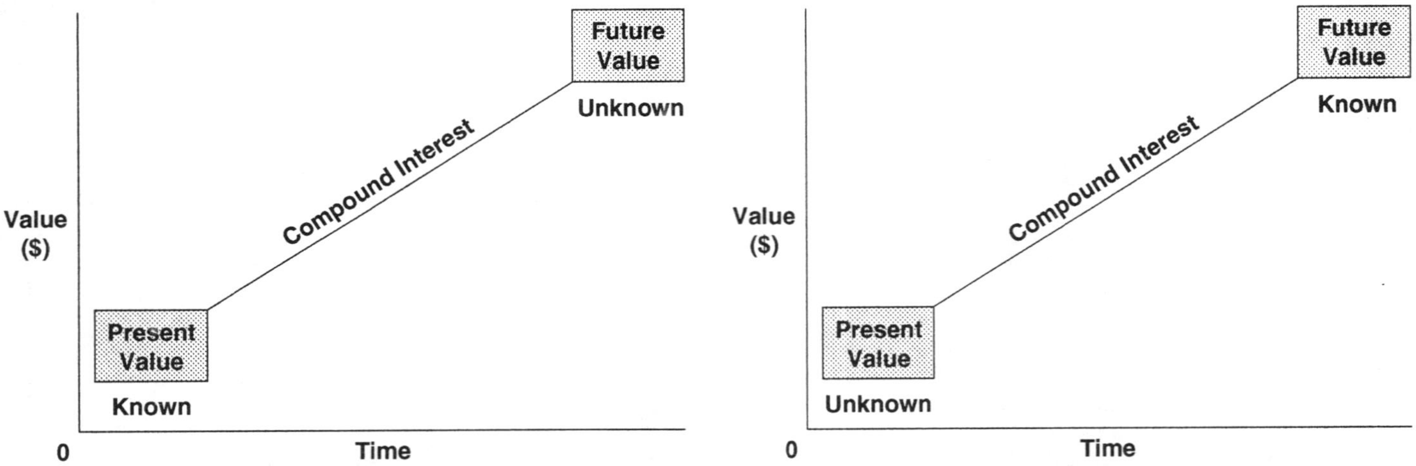

Compound Amount (Future Value) at Compound Interest

Present Amount (Present Value) at Compound Interest

TABLE 11-1

COMPOUND INTEREST TABLE (FUTURE VALUE OF $1 AT COMPOUND INTEREST)

Periods	$\frac{1}{2}$%	1%	$1\frac{1}{2}$%	2%	3%	4%	5%	6%	7%	8%	Periods
1	1.00500	1.01000	1.01500	1.02000	1.03000	1.04000	1.05000	1.06000	1.07000	1.08000	1
2	1.01003	1.02010	1.03023	1.04040	1.06090	1.08160	1.10250	1.12360	1.14490	1.16640	2
3	1.01508	1.03030	1.04568	1.06121	1.09273	1.12486	1.15763	1.19102	1.22504	1.25971	3
4	1.02015	1.04060	1.06136	1.08243	1.12551	1.16986	1.21551	1.26248	1.31080	1.36049	4
5	1.02525	1.05101	1.07728	1.10408	1.15927	1.21665	1.27628	1.33823	1.40255	1.46933	5
6	1.03038	1.06152	1.09344	1.12616	1.19405	1.26532	1.34010	1.41852	1.50073	1.58687	6
7	1.03553	1.07214	1.10984	1.14869	1.22987	1.31593	1.40710	1.50363	1.60578	1.71382	7
8	1.04071	1.08286	1.12649	1.17166	1.26677	1.36857	1.47746	1.59385	1.71819	1.85093	8
9	1.04591	1.09369	1.14339	1.19509	1.30477	1.42331	1.55133	1.68948	1.83846	1.99900	9
10	1.05114	1.10462	1.16054	1.21899	1.34392	1.48024	1.62889	1.79085	1.96715	2.15892	10
11	1.05640	1.11567	1.17795	1.24337	1.38423	1.53945	1.71034	1.89830	2.10485	2.33164	11
12	1.06168	1.12683	1.19562	1.26824	1.42576	1.60103	1.79586	2.01220	2.25219	2.51817	12
13	1.06699	1.13809	1.21355	1.29361	1.46853	1.66507	1.88565	2.13293	2.40985	2.71962	13
14	1.07232	1.14947	1.23176	1.31948	1.51259	1.73168	1.97993	2.26090	2.57853	2.93719	14
15	1.07768	1.16097	1.25023	1.34587	1.55797	1.80094	2.07893	2.39656	2.75903	3.17217	15
16	1.08307	1.17258	1.26899	1.37279	1.60471	1.87298	2.18287	2.54035	2.95216	3.42594	16
17	1.08849	1.18430	1.28802	1.40024	1.65285	1.94790	2.29202	2.69277	3.15882	3.70002	17
18	1.09393	1.19615	1.30734	1.42825	1.70243	2.02582	2.40662	2.85434	3.37993	3.99602	18
19	1.09940	1.20811	1.32695	1.45681	1.75351	2.10685	2.52695	3.02560	3.61653	4.31570	19
20	1.10490	1.22019	1.34686	1.48595	1.80611	2.19112	2.65330	3.20714	3.86968	4.66096	20
21	1.11042	1.23239	1.36706	1.51567	1.86029	2.27877	2.78596	3.39956	4.14056	5.03383	21
22	1.11597	1.24472	1.38756	1.54598	1.91610	2.36992	2.92526	3.60354	4.43040	5.43654	22
23	1.12155	1.25716	1.40838	1.57690	1.97359	2.46472	3.07152	3.81975	4.74053	5.87146	23
24	1.12716	1.26973	1.42950	1.60844	2.03279	2.56330	3.22510	4.04893	5.07237	6.34118	24
25	1.13280	1.28243	1.45095	1.64061	2.09378	2.66584	3.38635	4.29187	5.42743	6.84848	25

Periods	9%	10%	11%	12%	13%	14%	15%	16%	17%	18%	Periods
1	1.09000	1.10000	1.11000	1.12000	1.13000	1.14000	1.15000	1.16000	1.17000	1.18000	1
2	1.18810	1.21000	1.23210	1.25440	1.27690	1.29960	1.32250	1.34560	1.36890	1.39240	2
3	1.29503	1.33100	1.36763	1.40493	1.44290	1.48154	1.52088	1.56090	1.60161	1.64303	3
4	1.41158	1.46410	1.51807	1.57352	1.63047	1.68896	1.74901	1.81064	1.87389	1.93878	4
5	1.53862	1.61051	1.68506	1.76234	1.84244	1.92541	2.01136	2.10034	2.19245	2.28776	5
6	1.67710	1.77156	1.87041	1.97382	2.08195	2.19497	2.31306	2.43640	2.56516	2.69955	6
7	1.82804	1.94872	2.07616	2.21068	2.35261	2.50227	2.66002	2.82622	3.00124	3.18547	7
8	1.99256	2.14359	2.30454	2.47596	2.65844	2.85259	3.05902	3.27841	3.51145	3.75886	8
9	2.17189	2.35795	2.55804	2.77308	3.00404	3.25195	3.51788	3.80296	4.10840	4.43545	9
10	2.36736	2.59374	2.83942	3.10585	3.39457	3.70722	4.04556	4.41144	4.80683	5.23384	10
11	2.58043	2.85312	3.15176	3.47855	3.83586	4.22623	4.65239	5.11726	5.62399	6.17593	11
12	2.81266	3.13843	3.49845	3.89598	4.33452	4.81790	5.35025	5.93603	6.58007	7.28759	12
13	3.06580	3.45227	3.88328	4.36349	4.89801	5.49241	6.15279	6.88579	7.69868	8.59936	13
14	3.34173	3.79750	4.31044	4.88711	5.53475	6.26135	7.07571	7.98752	9.00745	10.14724	14
15	3.64248	4.17725	4.78459	5.47357	6.25427	7.13794	8.13706	9.26552	10.53872	11.97375	15
16	3.97031	4.59497	5.31089	6.13039	7.06733	8.13725	9.35762	10.74800	12.33030	14.12902	16
17	4.32763	5.05447	5.89509	6.86604	7.98608	9.27646	10.76126	12.46768	14.42646	16.67225	17
18	4.71712	5.55992	6.54355	7.68997	9.02427	10.57517	12.37545	14.46251	16.87895	19.67325	18
19	5.14166	6.11591	7.26334	8.61276	10.19742	12.05569	14.23177	16.77652	19.74838	23.21444	19
20	5.60441	6.72750	8.06231	9.64629	11.52309	13.74349	16.36654	19.46076	23.10560	27.39303	20
21	6.10881	7.40025	8.94917	10.80385	13.02109	15.66758	18.82152	22.57448	27.03355	32.32378	21
22	6.65860	8.14027	9.93357	12.10031	14.71383	17.86104	21.64475	26.18640	31.62925	38.14206	22
23	7.25787	8.95430	11.02627	13.55235	16.62663	20.36158	24.89146	30.37622	37.00623	45.00763	23
24	7.91108	9.84973	12.23916	15.17863	18.78809	23.21221	28.62518	35.23642	43.29729	53.10901	24
25	8.62308	10.83471	13.58546	17.00006	21.23054	26.46192	32.91895	40.87424	50.65783	62.66863	25

TABLE 11-2

PRESENT VALUE TABLE (PRESENT VALUE OF $1 AT COMPOUND INTEREST)

Periods	$\frac{1}{2}$%	1%	$1\frac{1}{2}$%	2%	3%	4%	5%	6%	7%	8%	Periods
1	0.99502	0.99010	0.98522	0.98039	0.97087	0.96154	0.95238	0.94340	0.93458	0.92593	1
2	0.99007	0.98030	0.97066	0.96117	0.94260	0.92456	0.90703	0.89000	0.87344	0.85734	2
3	0.98515	0.97059	0.95632	0.94232	0.91514	0.88900	0.86384	0.83962	0.81630	0.79383	3
4	0.98025	0.96098	0.94218	0.92385	0.88849	0.85480	0.82270	0.79209	0.76290	0.73503	4
5	0.97537	0.95147	0.92826	0.90573	0.86261	0.82193	0.78353	0.74726	0.71299	0.68058	5
6	0.97052	0.94205	0.91454	0.88797	0.83748	0.79031	0.74622	0.70496	0.66634	0.63017	6
7	0.96569	0.93272	0.90103	0.87056	0.81309	0.75992	0.71068	0.66506	0.62275	0.58349	7
8	0.96089	0.92348	0.88771	0.85349	0.78941	0.73069	0.67684	0.62741	0.58201	0.54027	8
9	0.95610	0.91434	0.87459	0.83676	0.76642	0.70259	0.64461	0.59190	0.54393	0.50025	9
10	0.95135	0.90529	0.86167	0.82035	0.74409	0.67556	0.61391	0.55839	0.50835	0.46319	10
11	0.94661	0.89632	0.84893	0.80426	0.72242	0.64958	0.58468	0.52679	0.47509	0.42888	11
12	0.94191	0.88745	0.83639	0.78849	0.70138	0.62460	0.55684	0.49697	0.44401	0.39711	12
13	0.93722	0.87866	0.82403	0.77303	0.68095	0.60057	0.53032	0.46884	0.41496	0.36770	13
14	0.93256	0.86996	0.81185	0.75788	0.66112	0.57748	0.50507	0.44230	0.38782	0.34046	14
15	0.92792	0.86135	0.79985	0.74301	0.64186	0.55526	0.48102	0.41727	0.36245	0.31524	15
16	0.92330	0.85282	0.78803	0.72845	0.62317	0.53391	0.45811	0.39365	0.33873	0.29189	16
17	0.91871	0.84438	0.77639	0.71416	0.60502	0.51337	0.43630	0.37136	0.31657	0.27027	17
18	0.91414	0.83602	0.76491	0.70016	0.58739	0.49363	0.41552	0.35034	0.29586	0.25025	18
19	0.90959	0.82774	0.75361	0.68643	0.57029	0.47464	0.39573	0.33051	0.27651	0.23171	19
20	0.90506	0.81954	0.74247	0.67297	0.55368	0.45639	0.37689	0.31180	0.25842	0.21455	20
21	0.90056	0.81143	0.73150	0.65978	0.53755	0.43883	0.35894	0.29416	0.24151	0.19866	21
22	0.89608	0.80340	0.72069	0.64684	0.52189	0.42196	0.34185	0.27751	0.22571	0.18394	22
23	0.89162	0.79544	0.71004	0.63416	0.50669	0.40573	0.32557	0.26180	0.21095	0.17032	23
24	0.88719	0.78757	0.69954	0.62172	0.49193	0.39012	0.31007	0.24698	0.19715	0.15770	24
25	0.88277	0.77977	0.68921	0.60953	0.47761	0.37512	0.29530	0.23300	0.18425	0.14602	25

Periods	9%	10%	11%	12%	13%	14%	15%	16%	17%	18%	Periods
1	0.91743	0.90909	0.90090	0.89286	0.88496	0.87719	0.86957	0.86207	0.85470	0.84746	1
2	0.84168	0.82645	0.81162	0.79719	0.78315	0.76947	0.75614	0.74316	0.73051	0.71818	2
3	0.77218	0.75131	0.73119	0.71178	0.69305	0.67497	0.65752	0.64066	0.62437	0.60863	3
4	0.70843	0.68301	0.65873	0.63552	0.61332	0.59208	0.57175	0.55229	0.53365	0.51579	4
5	0.64993	0.62092	0.59345	0.56743	0.54276	0.51937	0.49718	0.47611	0.45611	0.43711	5
6	0.59627	0.56447	0.53464	0.50663	0.48032	0.45559	0.43233	0.41044	0.38984	0.37043	6
7	0.54703	0.51316	0.48166	0.45235	0.42506	0.39964	0.37594	0.35383	0.33320	0.31393	7
8	0.50187	0.46651	0.43393	0.40388	0.37616	0.35056	0.32690	0.30503	0.28478	0.26604	8
9	0.46043	0.42410	0.39092	0.36061	0.33288	0.30751	0.28426	0.26295	0.24340	0.22546	9
10	0.42241	0.38554	0.35218	0.32197	0.29459	0.26974	0.24718	0.22668	0.20804	0.19106	10
11	0.38753	0.35049	0.31728	0.28748	0.26070	0.23662	0.21494	0.19542	0.17781	0.16192	11
12	0.35553	0.31863	0.28584	0.25668	0.23071	0.20756	0.18691	0.16846	0.15197	0.13722	12
13	0.32618	0.28966	0.25751	0.22917	0.20416	0.18207	0.16253	0.14523	0.12989	0.11629	13
14	0.29925	0.26333	0.23199	0.20462	0.18068	0.15971	0.14133	0.12520	0.11102	0.09855	14
15	0.27454	0.23939	0.20900	0.18270	0.15989	0.14010	0.12289	0.10793	0.09489	0.08352	15
16	0.25187	0.21763	0.18829	0.16312	0.14150	0.12289	0.10686	0.09304	0.08110	0.07078	16
17	0.23107	0.19784	0.16963	0.14564	0.12522	0.10780	0.09293	0.08021	0.06932	0.05998	17
18	0.21199	0.17986	0.15282	0.13004	0.11081	0.09456	0.08081	0.06914	0.05925	0.05083	18
19	0.19449	0.16351	0.13768	0.11611	0.09806	0.08295	0.07027	0.05961	0.05064	0.04308	19
20	0.17843	0.14864	0.12403	0.10367	0.08678	0.07276	0.06110	0.05139	0.04328	0.03651	20
21	0.16370	0.13513	0.11174	0.09256	0.07680	0.06383	0.05313	0.04430	0.03699	0.03094	21
22	0.15018	0.12285	0.10067	0.08264	0.06796	0.05599	0.04620	0.03819	0.03162	0.02622	22
23	0.13778	0.11168	0.09069	0.07379	0.06014	0.04911	0.04017	0.03292	0.02702	0.02222	23
24	0.12640	0.10153	0.08170	0.06588	0.05323	0.04308	0.03493	0.02838	0.02310	0.01883	24
25	0.11597	0.09230	0.07361	0.05882	0.04710	0.03779	0.03038	0.02447	0.01974	0.01596	25

EXHIBIT 12-1

TIME LINE ILLUSTRATING PRESENT AND FUTURE VALUE OF ANNUITIES

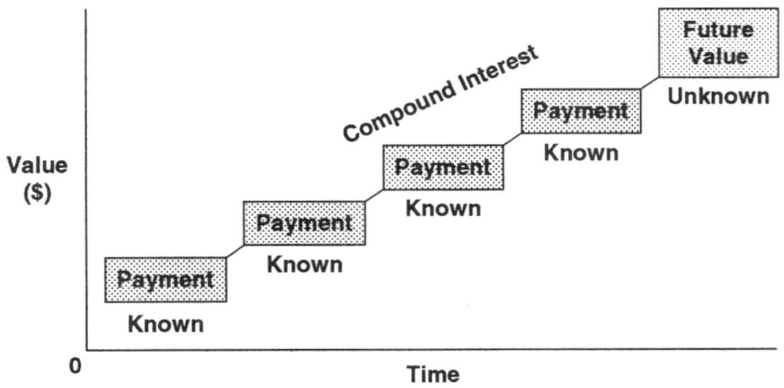

Future Value of an Annuity

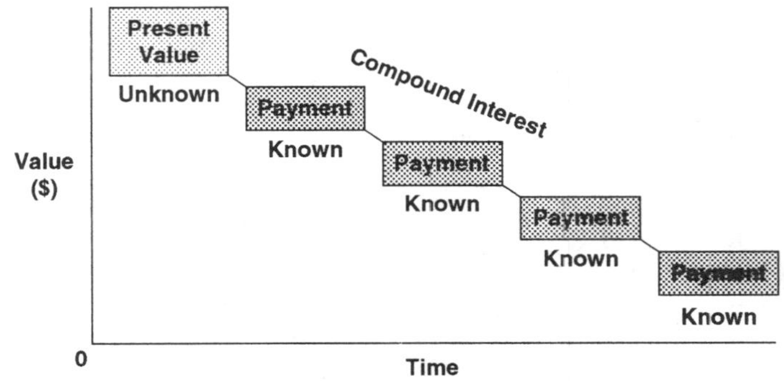

Present Value of an Annuity

TABLE 12-1

FUTURE VALUE (AMOUNT) OF AN ORDINARY ANNUITY OF $1.00

Periods	$\frac{1}{2}$%	1%	$1\frac{1}{2}$%	2%	3%	4%	5%	6%	7%	8%	Periods
1	1.00000	1.00000	1.00000	1.00000	1.00000	1.00000	1.00000	1.00000	1.00000	1.00000	1
2	2.00500	2.01000	2.01500	2.02000	2.03000	2.04000	2.05000	2.06000	2.07000	2.08000	2
3	3.01502	3.03010	3.04522	3.06040	3.09090	3.12160	3.15250	3.18360	3.21490	3.24640	3
4	4.03010	4.06040	4.09090	4.12161	4.18363	4.24646	4.31013	4.37462	4.43994	4.50611	4
5	5.05025	5.10101	5.15227	5.20404	5.30914	5.41632	5.52563	5.63709	5.75074	5.86660	5
6	6.07550	6.15202	6.22955	6.30812	6.46841	6.63298	6.80191	6.97532	7.15329	7.33593	6
7	7.10588	7.21354	7.32299	7.43428	7.66246	7.89829	8.14201	8.39384	8.65402	8.92280	7
8	8.14141	8.28567	8.43284	8.58297	8.89234	9.21423	9.54911	9.89747	10.25980	10.63663	8
9	9.18212	9.36853	9.55933	9.75463	10.15911	10.58280	11.02656	11.49132	11.97799	12.48756	9
10	10.22803	10.46221	10.70272	10.94972	11.46388	12.00611	12.57789	13.18079	13.81645	14.48656	10
11	11.27917	11.56683	11.86326	12.16872	12.80780	13.48635	14.20679	14.97164	15.78360	16.64549	11
12	12.33556	12.68250	13.04121	13.41209	14.19203	15.02581	15.91713	16.86994	17.88845	18.97713	12
13	13.39724	13.80933	14.23683	14.68033	15.61779	16.62684	17.71298	18.88214	20.14064	21.49530	13
14	14.46423	14.94742	15.45038	15.97394	17.08632	18.29191	19.59863	21.01507	22.55049	24.21492	14
15	15.53655	16.09690	16.68214	17.29342	18.59891	20.02359	21.57856	23.27597	25.12902	27.15211	15
16	16.61423	17.25786	17.93237	18.63929	20.15688	21.82453	23.65749	25.67253	27.88805	30.32428	16
17	17.69730	18.43044	19.20136	20.01207	21.76159	23.69751	25.84037	28.21288	30.84022	33.75023	17
18	18.78579	19.61475	20.48938	21.41231	23.41444	25.64541	28.13238	30.90565	33.99903	37.45024	18
19	19.87972	20.81090	21.79672	22.84056	25.11687	27.67123	30.53900	33.75999	37.37896	41.44626	19
20	20.97912	22.01900	23.12367	24.29737	26.87037	29.77808	33.06595	36.78559	40.99549	45.76196	20
21	22.08401	23.23919	24.47052	25.78332	28.67649	31.96920	35.71925	39.99273	44.86518	50.42292	21
22	23.19443	24.47159	25.83758	27.29898	30.53678	34.24797	38.50521	43.39229	49.00574	55.45676	22
23	24.31040	25.71630	27.22514	28.84496	32.45288	36.61789	41.43048	46.99583	53.43614	60.89330	23
24	25.43196	26.97346	28.63352	30.42186	34.42647	39.08260	44.50200	50.81558	58.17667	66.76476	24
25	26.55912	28.24320	30.06302	32.03030	36.45926	41.64591	47.72710	54.86451	63.24904	73.10594	25
26	27.69191	29.52563	31.51397	33.67091	38.55304	44.31174	51.11345	59.15638	68.67647	79.95442	26
27	28.83037	30.82089	32.98668	35.34432	40.70963	47.08421	54.66913	63.70577	74.48382	87.35077	27
28	29.97452	32.12910	34.48148	37.05121	42.93092	49.96758	58.40258	68.52811	80.69769	95.33883	28
29	31.12439	33.45039	35.99870	38.79223	45.21885	52.96629	62.32271	73.63980	87.34653	103.96594	29
30	32.28002	34.78489	37.53868	40.56808	47.57542	56.08494	66.43885	79.05819	94.46079	113.28321	30
31	33.44142	36.13274	39.10176	42.37944	50.00268	59.32834	70.76079	84.80168	102.07304	123.34587	31
32	34.60862	37.49407	40.68829	44.22703	52.50276	62.70147	75.29883	90.88978	110.21815	134.21354	32
33	35.78167	38.86901	42.29861	46.11157	55.07784	66.20953	80.06377	97.34316	118.93343	145.95062	33
34	36.96058	40.25770	43.93309	48.03380	57.73018	69.85791	85.06696	104.18375	128.25876	158.62667	34
35	38.14538	41.66028	45.59209	49.99448	60.46208	73.65222	90.32031	111.43478	138.23688	172.31680	35
36	39.33610	43.07688	47.27597	51.99437	63.27594	77.59831	95.83632	119.12087	148.91346	187.10215	36

Periods	9%	10%	11%	12%	13%	14%	15%	16%	17%	18%	Periods
1	1.00000	1.00000	1.00000	1.00000	1.00000	1.00000	1.00000	1.00000	1.00000	1.00000	1
2	2.09000	2.10000	2.11000	2.12000	2.13000	2.14000	2.15000	2.16000	2.17000	2.18000	2
3	3.27810	3.31000	3.34210	3.37440	3.40690	3.43960	3.47250	3.50560	3.53890	3.57240	3
4	4.57313	4.64100	4.70973	4.77933	4.84980	4.92114	4.99338	5.06650	5.14051	5.21543	4
5	5.98471	6.10510	6.22780	6.35285	6.48027	6.61010	6.74238	6.87714	7.01440	7.15421	5
6	7.52333	7.71561	7.91286	8.11519	8.32271	8.53552	8.75374	8.97748	9.20685	9.44197	6
7	9.20043	9.48717	9.78327	10.08901	10.40466	10.73049	11.06680	11.41387	11.77201	12.14152	7
8	11.02847	11.43589	11.85943	12.29969	12.75726	13.23276	13.72682	14.24009	14.77325	15.32700	8
9	13.02104	13.57948	14.16397	14.77566	15.41571	16.08535	16.78584	17.51851	18.28471	19.08585	9
10	15.19293	15.93742	16.72201	17.54874	18.41975	19.33730	20.30372	21.32147	22.39311	23.52131	10
11	17.56029	18.53117	19.56143	20.65458	21.81432	23.04452	24.34928	25.73290	27.19994	28.75514	11
12	20.14072	21.38428	22.71319	24.13313	25.65018	27.27075	29.00167	30.85017	32.82393	34.93107	12
13	22.95338	24.52271	26.21164	28.02911	29.98470	32.08865	34.35192	36.78620	39.40399	42.21866	13
14	26.01919	27.97498	30.09492	32.39260	34.88271	37.58107	40.50471	43.67199	47.10267	50.81802	14
15	29.36092	31.77248	34.40536	37.27971	40.41746	43.84241	47.58041	51.65951	56.11013	60.96527	15
16	33.00340	35.94973	39.18995	42.75328	46.67173	50.98035	55.71747	60.92503	66.64885	72.93901	16
17	36.97370	40.54470	44.50084	48.88367	53.73906	59.11760	65.07509	71.67303	78.97915	87.06804	17
18	41.30134	45.59917	50.39594	55.74971	61.72514	68.39407	75.83636	84.14072	93.40561	103.74028	18
19	46.01846	51.15909	56.93949	63.43968	70.74941	78.96923	88.21181	98.60323	110.28456	123.41353	19
20	51.16012	57.27500	64.20283	72.05244	80.94683	91.02493	102.44358	115.37975	130.03294	146.62797	20
21	56.76453	64.00250	72.26514	81.69874	92.46992	104.76842	118.81012	134.84051	153.13854	174.02100	21
22	62.87334	71.40275	81.21431	92.50258	105.49101	120.43600	137.63164	157.41499	180.17209	206.34479	22
23	69.53194	79.54302	91.14788	104.60289	120.20484	138.29704	159.27638	183.60138	211.80134	244.48685	23
24	76.78981	88.49733	102.17415	118.15524	136.83147	158.65862	184.16784	213.97761	248.80757	289.49448	24
25	84.70090	98.34706	114.41331	133.33387	155.61956	181.87083	212.79302	249.21402	292.10486	342.60349	25
26	93.32398	109.18177	127.99877	150.33393	176.85010	208.33274	245.71197	290.08827	342.76268	405.27211	26
27	102.72313	121.09994	143.07864	169.37401	200.84061	238.49933	283.56877	337.50239	402.03234	479.22109	27
28	112.96822	134.20994	159.81729	190.69889	227.94989	272.88923	327.10408	392.50277	471.37783	566.48089	28
29	124.13536	148.63093	178.39719	214.58275	258.58338	312.09373	377.16969	456.30322	552.51207	669.44745	29
30	136.30754	164.49402	199.02088	241.33268	293.19922	356.78685	434.74515	530.31173	647.43912	790.94799	30
31	149.57522	181.94342	221.91317	271.29261	332.31511	407.73701	500.95692	616.16161	758.50377	934.31863	31
32	164.03699	201.13777	247.32362	304.84772	376.51608	465.82019	577.10046	715.74746	888.44941	1103.49598	32
33	179.80032	222.25154	275.52922	342.42945	426.46317	532.03501	664.66552	831.26706	1040.48581	1303.12526	33
34	196.98234	245.47670	306.83744	384.52098	482.90338	607.51991	765.36535	965.26979	1218.36839	1538.68781	34
35	215.71075	271.02437	341.58955	431.66350	546.68082	693.57270	881.17016	1120.71295	1426.49102	1816.65161	35
36	236.12472	299.12681	380.16441	484.46312	618.74933	791.67288	1014.34568	1301.02703	1669.99450	2144.64890	36

TABLE 12-2

PRESENT VALUE (AMOUNT) OF AN ORDINARY ANNUITY OF $1.00

Periods	$\frac{1}{2}$%	1%	$1\frac{1}{2}$%	2%	3%	4%	5%	6%	7%	8%	Periods
1	0.99502	0.99010	0.98522	0.98039	0.97087	0.96154	0.95238	0.94340	0.93458	0.92593	1
2	1.98510	1.97040	1.95588	1.94156	1.91347	1.88609	1.85941	1.83339	1.80802	1.78326	2
3	2.97025	2.94099	2.91220	2.88388	2.82861	2.77509	2.72325	2.67301	2.62432	2.57710	3
4	3.95050	3.90197	3.85438	3.80773	3.71710	3.62990	3.54595	3.46511	3.38721	3.31213	4
5	4.92587	4.85343	4.78264	4.71346	4.57971	4.45182	4.32948	4.21236	4.10020	3.99271	5
6	5.89638	5.79548	5.69719	5.60143	5.41719	5.24214	5.07569	4.91732	4.76654	4.62288	6
7	6.86207	6.72819	6.59821	6.47199	6.23028	6.00205	5.78637	5.58238	5.38929	5.20637	7
8	7.82296	7.65168	7.48593	7.32548	7.01969	6.73274	6.46321	6.20979	5.97130	5.74664	8
9	8.77906	8.56602	8.36052	8.16224	7.78611	7.43533	7.10782	6.80169	6.51523	6.24689	9
10	9.73041	9.47130	9.22218	8.98259	8.53020	8.11090	7.72173	7.36009	7.02358	6.71008	10
11	10.67703	10.36763	10.07112	9.78685	9.25262	8.76048	8.30641	7.88687	7.49867	7.13896	11
12	11.61893	11.25508	10.90751	10.57534	9.95400	9.38507	8.86325	8.38384	7.94269	7.53608	12
13	12.55615	12.13374	11.73153	11.34837	10.63496	9.98565	9.39357	8.85268	8.35765	7.90378	13
14	13.48871	13.00370	12.54338	12.10625	11.29607	10.56312	9.89864	9.29498	8.74547	8.24424	14
15	14.41662	13.86505	13.34323	12.84926	11.93794	11.11839	10.37966	9.71225	9.10791	8.55948	15
16	15.33993	14.71787	14.13126	13.57771	12.56110	11.65230	10.83777	10.10590	9.44665	8.85137	16
17	16.25863	15.56225	14.90765	14.29187	13.16612	12.16567	11.27407	10.47726	9.76322	9.12164	17
18	17.17277	16.39827	15.67256	14.99203	13.75351	12.65930	11.68959	10.82760	10.05909	9.37189	18
19	18.08236	17.22601	16.42617	15.67846	14.32380	13.13394	12.08532	11.15812	10.33560	9.60360	19
20	18.98742	18.04555	17.16864	16.35143	14.87747	13.59033	12.46221	11.46992	10.59401	9.81815	20
21	19.88798	18.85698	17.90014	17.01121	15.41502	14.02916	12.82115	11.76408	10.83553	10.01680	21
22	20.78406	19.66038	18.62082	17.65805	15.93692	14.45112	13.16300	12.04158	11.06124	10.20074	22
23	21.67568	20.45582	19.33086	18.29220	16.44361	14.85684	13.48857	12.30338	11.27219	10.37106	23
24	22.56287	21.24339	20.03041	18.91393	16.93554	15.24696	13.79864	12.55036	11.46933	10.52876	24
25	23.44564	22.02316	20.71961	19.52346	17.41315	15.62208	14.09394	12.78336	11.65358	10.67478	25
26	24.32402	22.79520	21.39863	20.12104	17.87684	15.98277	14.37519	13.00317	11.82578	10.80998	26
27	25.19803	23.55961	22.06762	20.70690	18.32703	16.32959	14.64303	13.21053	11.98671	10.93516	27
28	26.06769	24.31644	22.72672	21.28127	18.76411	16.66306	14.89813	13.40616	12.13711	11.05108	28
29	26.93302	25.06579	23.37608	21.84438	19.18845	16.98371	15.14107	13.59072	12.27767	11.15841	29
30	27.79405	25.80771	24.01584	22.39646	19.60044	17.29203	15.37245	13.76483	12.40904	11.25778	30
31	28.65080	26.54229	24.64615	22.93770	20.00043	17.58849	15.59281	13.92909	12.53181	11.34980	31
32	29.50328	27.26959	25.26714	23.46833	20.38877	17.87355	15.80268	14.08404	12.64656	11.43500	32
33	30.35153	27.98969	25.87895	23.98856	20.76579	18.14765	16.00255	14.23023	12.75379	11.51389	33
34	31.19555	28.70267	26.48173	24.49859	21.13184	18.41120	16.19290	14.36814	12.85401	11.58693	34
35	32.03537	29.40858	27.07559	24.99862	21.48722	18.66461	16.37419	14.49825	12.94767	11.65457	35
36	32.87102	30.10751	27.66068	25.48884	21.83225	18.90828	16.54685	14.62099	13.03521	11.71719	36

Periods	9%	10%	11%	12%	13%	14%	15%	16%	17%	18%	Periods
1	0.91743	0.90909	0.90090	0.89286	0.88496	0.87719	0.86957	0.86207	0.85470	0.84746	1
2	1.75911	1.73554	1.71252	1.69005	1.66810	1.64666	1.62571	1.60523	1.58521	1.56564	2
3	2.53129	2.48685	2.44371	2.40183	2.36115	2.32163	2.28323	2.24589	2.20958	2.17427	3
4	3.23972	3.16987	3.10245	3.03735	2.97447	2.91371	2.85498	2.79818	2.74324	2.69006	4
5	3.88965	3.79079	3.69590	3.60478	3.51723	3.43308	3.35216	3.27429	3.19935	3.12717	5
6	4.48592	4.35526	4.23054	4.11141	3.99755	3.88867	3.78448	3.68474	3.58918	3.49760	6
7	5.03295	4.86842	4.71220	4.56376	4.42261	4.28830	4.16042	4.03857	3.92238	3.81153	7
8	5.53482	5.33493	5.14612	4.96764	4.79877	4.63886	4.48732	4.34359	4.20716	4.07757	8
9	5.99525	5.75902	5.53705	5.32825	5.13166	4.94637	4.77158	4.60654	4.45057	4.30302	9
10	6.41766	6.14457	5.88923	5.65022	5.42624	5.21612	5.01877	4.83323	4.65860	4.49409	10
11	6.80519	6.49506	6.20652	5.93770	5.68694	5.45273	5.23371	5.02864	4.83641	4.65601	11
12	7.16073	6.81369	6.49236	6.19437	5.91765	5.66029	5.42062	5.19711	4.98839	4.79322	12
13	7.48690	7.10336	6.74987	6.42355	6.12181	5.84236	5.58315	5.34233	5.11828	4.90951	13
14	7.78615	7.36669	6.98187	6.62817	6.30249	6.00207	5.72448	5.46753	5.22930	5.00806	14
15	8.06069	7.60608	7.19087	6.81086	6.46238	6.14217	5.84737	5.57546	5.32419	5.09158	15
16	8.31256	7.82371	7.37916	6.97399	6.60388	6.26506	5.95423	5.66850	5.40529	5.16235	16
17	8.54363	8.02155	7.54879	7.11963	6.72909	6.37286	6.04716	5.74870	5.47461	5.22233	17
18	8.75563	8.20141	7.70162	7.24967	6.83991	6.46742	6.12797	5.81785	5.53385	5.27316	18
19	8.95011	8.36492	7.83929	7.36578	6.93797	6.55037	6.19823	5.87746	5.58449	5.31624	19
20	9.12855	8.51356	7.96333	7.46944	7.02475	6.62313	6.25933	5.92884	5.62777	5.35275	20
21	9.29224	8.64869	8.07507	7.56200	7.10155	6.68696	6.31246	5.97314	5.66476	5.38368	21
22	9.44243	8.77154	8.17574	7.64465	7.16951	6.74294	6.35866	6.01133	5.69637	5.40990	22
23	9.58021	8.88322	8.26643	7.71843	7.22966	6.79206	6.39884	6.04425	5.72340	5.43212	23
24	9.70661	8.98474	8.34814	7.78432	7.28288	6.83514	6.43377	6.07263	5.74649	5.45095	24
25	9.82258	9.07704	8.42174	7.84314	7.32998	6.87293	6.46415	6.09709	5.76623	5.46691	25
26	9.92897	9.16095	8.48806	7.89566	7.37167	6.90608	6.49056	6.11818	5.78311	5.48043	26
27	10.02658	9.23722	8.54780	7.94255	7.40856	6.93515	6.51353	6.13636	5.79753	5.49189	27
28	10.11613	9.30657	8.60162	7.98442	7.44120	6.96066	6.53351	6.15204	5.80985	5.50160	28
29	10.19828	9.36961	8.65011	8.02181	7.47009	6.98304	6.55088	6.16555	5.82039	5.50983	29
30	10.27365	9.42691	8.69379	8.05518	7.49565	7.00266	6.56598	6.17720	5.82939	5.51681	30
31	10.34280	9.47901	8.73315	8.08499	7.51828	7.01988	6.57911	6.18724	5.83709	5.52272	31
32	10.40624	9.52638	8.76860	8.11159	7.53830	7.03498	6.59053	6.19590	5.84366	5.52773	32
33	10.46444	9.56943	8.80054	8.13535	7.55602	7.04823	6.60046	6.20336	5.84928	5.53197	33
34	10.51784	9.60857	8.82932	8.15656	7.57170	7.05985	6.60910	6.20979	5.85409	5.53557	34
35	10.56682	9.64416	8.85524	8.17550	7.58557	7.07005	6.61661	6.21534	5.85820	5.53862	35
36	10.61176	9.67651	8.87859	8.19241	7.59785	7.07899	6.62314	6.22012	5.86171	5.54120	36

TABLE 13-1

ANNUAL PERCENTAGE RATE (APR) FINANCE CHARGE PER $100

ANNUAL PERCENTAGE RATE TABLE FOR MONTHLY PAYMENT PLANS
SEE INSTRUCTIONS FOR USE OF TABLES

FRB-103-M

NUMBER OF PAYMENTS	ANNUAL PERCENTAGE RATE															
	10.00%	10.25%	10.50%	10.75%	11.00%	11.25%	11.50%	11.75%	12.00%	12.25%	12.50%	12.75%	13.00%	13.25%	13.50%	13.75%
	(FINANCE CHARGE PER $100 OF AMOUNT FINANCED)															
1	0.83	0.85	0.87	0.90	0.92	0.94	0.96	0.98	1.00	1.02	1.04	1.06	1.08	1.10	1.12	1.15
2	1.25	1.28	1.31	1.35	1.38	1.41	1.44	1.47	1.50	1.53	1.57	1.60	1.63	1.66	1.69	1.72
3	1.67	1.71	1.76	1.80	1.84	1.88	1.92	1.96	2.01	2.05	2.09	2.13	2.17	2.22	2.26	2.30
4	2.09	2.14	2.20	2.25	2.30	2.35	2.41	2.46	2.51	2.57	2.62	2.67	2.72	2.78	2.83	2.88
5	2.51	2.58	2.64	2.70	2.77	2.83	2.89	2.96	3.02	3.08	3.15	3.21	3.27	3.34	3.40	3.46
6	2.94	3.01	3.08	3.16	3.23	3.31	3.38	3.45	3.53	3.60	3.68	3.75	3.83	3.90	3.97	4.05
7	3.36	3.45	3.53	3.62	3.70	3.78	3.87	3.95	4.04	4.12	4.21	4.29	4.38	4.47	4.55	4.64
8	3.79	3.88	3.98	4.07	4.17	4.26	4.36	4.46	4.55	4.65	4.74	4.84	4.94	5.03	5.13	5.22
9	4.21	4.32	4.43	4.53	4.64	4.75	4.85	4.96	5.07	5.17	5.28	5.39	5.49	5.60	5.71	5.82
10	4.64	4.76	4.88	4.99	5.11	5.23	5.35	5.46	5.58	5.70	5.82	5.94	6.05	6.17	6.29	6.41
11	5.07	5.20	5.33	5.45	5.58	5.71	5.84	5.97	6.10	6.23	6.36	6.49	6.62	6.75	6.88	7.01
12	5.50	5.64	5.78	5.92	6.06	6.20	6.34	6.48	6.62	6.76	6.90	7.04	7.18	7.32	7.46	7.60
13	5.93	6.08	6.23	6.38	6.53	6.68	6.84	6.99	7.14	7.29	7.44	7.59	7.75	7.90	8.05	8.20
14	6.36	6.52	6.69	6.85	7.01	7.17	7.34	7.50	7.66	7.82	7.99	8.15	8.31	8.48	8.64	8.81
15	6.80	6.97	7.14	7.32	7.49	7.66	7.84	8.01	8.19	8.36	8.53	8.71	8.88	9.06	9.23	9.41
16	7.23	7.41	7.60	7.78	7.97	8.15	8.34	8.53	8.71	8.90	9.08	9.27	9.46	9.64	9.83	10.02
17	7.67	7.86	8.06	8.25	8.45	8.65	8.84	9.04	9.24	9.44	9.63	9.83	10.03	10.23	10.43	10.63
18	8.10	8.31	8.52	8.73	8.93	9.14	9.35	9.56	9.77	9.98	10.19	10.40	10.61	10.82	11.03	11.24
19	8.54	8.76	8.98	9.20	9.42	9.64	9.86	10.08	10.30	10.52	10.74	10.96	11.18	11.41	11.63	11.85
20	8.98	9.21	9.44	9.67	9.90	10.13	10.37	10.60	10.83	11.06	11.30	11.53	11.76	12.00	12.23	12.46
21	9.42	9.66	9.90	10.15	10.39	10.63	10.88	11.12	11.36	11.61	11.85	12.10	12.34	12.59	12.84	13.08
22	9.86	10.12	10.37	10.62	10.88	11.13	11.39	11.64	11.90	12.16	12.41	12.67	12.93	13.19	13.44	13.70
23	10.30	10.57	10.84	11.10	11.37	11.63	11.90	12.17	12.44	12.71	12.97	13.24	13.51	13.78	14.05	14.32
24	10.75	11.02	11.30	11.58	11.86	12.14	12.42	12.70	12.98	13.26	13.54	13.82	14.10	14.38	14.66	14.95
25	11.19	11.48	11.77	12.06	12.35	12.64	12.93	13.22	13.52	13.81	14.10	14.40	14.69	14.98	15.28	15.57
26	11.64	11.94	12.24	12.54	12.85	13.15	13.45	13.75	14.06	14.36	14.67	14.97	15.28	15.59	15.89	16.20
27	12.09	12.40	12.71	13.03	13.34	13.66	13.97	14.29	14.60	14.92	15.24	15.56	15.87	16.19	16.51	16.83
28	12.53	12.86	13.18	13.51	13.84	14.16	14.49	14.82	15.15	15.48	15.81	16.14	16.47	16.80	17.13	17.46
29	12.98	13.32	13.66	14.00	14.33	14.67	15.01	15.35	15.70	16.04	16.38	16.72	17.07	17.41	17.75	18.10
30	13.43	13.78	14.13	14.48	14.83	15.19	15.54	15.89	16.24	16.60	16.95	17.31	17.66	18.02	18.38	18.74
31	13.89	14.25	14.61	14.97	15.33	15.70	16.06	16.43	16.79	17.16	17.53	17.90	18.27	18.63	19.00	19.38
32	14.34	14.71	15.09	15.46	15.84	16.21	16.59	16.97	17.35	17.73	18.11	18.49	18.87	19.25	19.63	20.02
33	14.79	15.18	15.57	15.95	16.34	16.73	17.12	17.51	17.90	18.29	18.69	19.08	19.47	19.87	20.26	20.66
34	15.25	15.65	16.05	16.44	16.85	17.25	17.65	18.05	18.46	18.86	19.27	19.67	20.08	20.49	20.90	21.31
35	15.70	16.11	16.53	16.94	17.35	17.77	18.18	18.60	19.01	19.43	19.85	20.27	20.69	21.11	21.53	21.95
36	16.16	16.58	17.01	17.43	17.86	18.29	18.71	19.14	19.57	20.00	20.43	20.87	21.30	21.73	22.17	22.60
37	16.62	17.06	17.49	17.93	18.37	18.81	19.25	19.69	20.13	20.58	21.02	21.46	21.91	22.36	22.81	23.25
38	17.08	17.53	17.98	18.43	18.88	19.33	19.78	20.24	20.69	21.15	21.61	22.07	22.52	22.99	23.45	23.91
39	17.54	18.00	18.46	18.93	19.39	19.86	20.32	20.79	21.26	21.73	22.20	22.67	23.14	23.61	24.09	24.56
40	18.00	18.48	18.95	19.43	19.90	20.38	20.86	21.34	21.82	22.30	22.79	23.27	23.76	24.25	24.73	25.22
41	18.47	18.95	19.44	19.93	20.42	20.91	21.40	21.89	22.39	22.88	23.38	23.88	24.38	24.88	25.38	25.88
42	18.93	19.43	19.93	20.43	20.93	21.44	21.94	22.45	22.96	23.47	23.98	24.49	25.00	25.51	26.03	26.55
43	19.40	19.91	20.42	20.94	21.45	21.97	22.49	23.01	23.53	24.05	24.57	25.10	25.62	26.15	26.68	27.21
44	19.86	20.39	20.91	21.44	21.97	22.50	23.03	23.57	24.10	24.64	25.17	25.71	26.25	26.79	27.33	27.88
45	20.33	20.87	21.41	21.95	22.49	23.03	23.58	24.12	24.67	25.22	25.77	26.32	26.88	27.43	27.99	28.55
46	20.80	21.35	21.90	22.46	23.01	23.57	24.13	24.69	25.25	25.81	26.37	26.94	27.51	28.08	28.65	29.22
47	21.27	21.83	22.40	22.97	23.53	24.10	24.68	25.25	25.82	26.40	26.98	27.56	28.14	28.72	29.31	29.89
48	21.74	22.32	22.90	23.48	24.06	24.64	25.23	25.81	26.40	26.99	27.58	28.18	28.77	29.37	29.97	30.57
49	22.21	22.80	23.39	23.99	24.58	25.18	25.78	26.38	26.98	27.59	28.19	28.80	29.41	30.02	30.63	31.24
50	22.69	23.29	23.89	24.50	25.11	25.72	26.33	26.95	27.56	28.18	28.80	29.42	30.04	30.67	31.29	31.92
51	23.16	23.78	24.40	25.02	25.64	26.26	26.89	27.52	28.15	28.78	29.41	30.05	30.68	31.32	31.96	32.60
52	23.64	24.27	24.90	25.53	26.17	26.81	27.45	28.09	28.73	29.38	30.02	30.67	31.32	31.98	32.63	33.29
53	24.11	24.76	25.40	26.05	26.70	27.35	28.00	28.66	29.32	29.98	30.64	31.30	31.97	32.63	33.30	33.97
54	24.59	25.25	25.91	26.57	27.23	27.90	28.56	29.23	29.91	30.58	31.25	31.93	32.61	33.29	33.98	34.66
55	25.07	25.74	26.41	27.09	27.77	28.44	29.13	29.81	30.50	31.18	31.87	32.56	33.26	33.95	34.65	35.35
56	25.55	26.23	26.92	27.61	28.30	28.99	29.69	30.39	31.09	31.79	32.49	33.20	33.91	34.62	35.33	36.04
57	26.03	26.73	27.43	28.13	28.84	29.54	30.25	30.97	31.68	32.39	33.11	33.83	34.56	35.28	36.01	36.74
58	26.51	27.23	27.94	28.66	29.37	30.10	30.82	31.55	32.27	33.00	33.74	34.47	35.21	35.95	36.69	37.43
59	27.00	27.72	28.45	29.18	29.91	30.65	31.39	32.13	32.87	33.61	34.36	35.11	35.86	36.62	37.37	38.13
60	27.48	28.22	28.96	29.71	30.45	31.20	31.96	32.71	33.47	34.23	34.99	35.75	36.52	37.29	38.06	38.83

continued

TABLE 13-1

ANNUAL PERCENTAGE RATE (APR)
FINANCE CHARGE PER $100

ANNUAL PERCENTAGE RATE TABLE FOR MONTHLY PAYMENT PLANS
SEE INSTRUCTIONS FOR USE OF TABLES

FRB-104-M

NUMBER OF PAYMENTS	ANNUAL PERCENTAGE RATE															
	14.00%	14.25%	14.50%	14.75%	15.00%	15.25%	15.50%	15.75%	16.00%	16.25%	16.50%	16.75%	17.00%	17.25%	17.50%	17.75%
	(FINANCE CHARGE PER $100 OF AMOUNT FINANCED)															
1	1.17	1.19	1.21	1.23	1.25	1.27	1.29	1.31	1.33	1.35	1.37	1.40	1.42	1.44	1.46	1.48
2	1.75	1.78	1.82	1.85	1.88	1.91	1.94	1.97	2.00	2.04	2.07	2.10	2.13	2.16	2.19	2.22
3	2.34	2.38	2.43	2.47	2.51	2.55	2.59	2.64	2.68	2.72	2.76	2.80	2.85	2.89	2.93	2.97
4	2.93	2.99	3.04	3.09	3.14	3.20	3.25	3.30	3.36	3.41	3.46	3.51	3.57	3.62	3.67	3.73
5	3.53	3.59	3.65	3.72	3.78	3.84	3.91	3.97	4.04	4.10	4.16	4.23	4.29	4.35	4.42	4.48
6	4.12	4.20	4.27	4.35	4.42	4.49	4.57	4.64	4.72	4.79	4.87	4.94	5.02	5.09	5.17	5.24
7	4.72	4.81	4.89	4.98	5.06	5.15	5.23	5.32	5.40	5.49	5.58	5.66	5.75	5.83	5.92	6.00
8	5.32	5.42	5.51	5.61	5.71	5.80	5.90	6.00	6.09	6.19	6.29	6.38	6.48	6.58	6.67	6.77
9	5.92	6.03	6.14	6.25	6.35	6.46	6.57	6.68	6.78	6.89	7.00	7.11	7.22	7.32	7.43	7.54
10	6.53	6.65	6.77	6.88	7.00	7.12	7.24	7.36	7.48	7.60	7.72	7.84	7.96	8.08	8.19	8.31
11	7.14	7.27	7.40	7.53	7.66	7.79	7.92	8.05	8.18	8.31	8.44	8.57	8.70	8.83	8.96	9.09
12	7.74	7.89	8.03	8.17	8.31	8.45	8.59	8.74	8.88	9.02	9.16	9.30	9.45	9.59	9.73	9.87
13	8.36	8.51	8.66	8.81	8.97	9.12	9.27	9.43	9.58	9.73	9.89	10.04	10.20	10.35	10.50	10.66
14	8.97	9.13	9.30	9.46	9.63	9.79	9.96	10.12	10.29	10.45	10.62	10.78	10.95	11.11	11.28	11.45
15	9.59	9.76	9.94	10.11	10.29	10.47	10.64	10.82	11.00	11.17	11.35	11.53	11.71	11.88	12.06	12.24
16	10.20	10.39	10.58	10.77	10.95	11.14	11.33	11.52	11.71	11.90	12.09	12.28	12.46	12.65	12.84	13.03
17	10.82	11.02	11.22	11.42	11.62	11.82	12.02	12.22	12.42	12.62	12.83	13.03	13.23	13.43	13.63	13.83
18	11.45	11.66	11.87	12.08	12.29	12.50	12.72	12.93	13.14	13.35	13.57	13.78	13.99	14.21	14.42	14.64
19	12.07	12.30	12.52	12.74	12.97	13.19	13.41	13.64	13.86	14.09	14.31	14.54	14.76	14.99	15.22	15.44
20	12.70	12.93	13.17	13.41	13.64	13.88	14.11	14.35	14.59	14.82	15.06	15.30	15.54	15.77	16.01	16.25
21	13.33	13.58	13.82	14.07	14.32	14.57	14.82	15.06	15.31	15.56	15.81	16.06	16.31	16.56	16.81	17.07
22	13.96	14.22	14.48	14.74	15.00	15.26	15.52	15.78	16.04	16.30	16.57	16.83	17.09	17.36	17.62	17.88
23	14.59	14.87	15.14	15.41	15.68	15.96	16.23	16.50	16.78	17.05	17.32	17.60	17.88	18.15	18.43	18.70
24	15.23	15.51	15.80	16.08	16.37	16.65	16.94	17.22	17.51	17.80	18.09	18.37	18.66	18.95	19.24	19.53
25	15.87	16.17	16.46	16.76	17.06	17.35	17.65	17.95	18.25	18.55	18.85	19.15	19.45	19.75	20.05	20.36
26	16.51	16.82	17.13	17.44	17.75	18.06	18.37	18.68	18.99	19.30	19.62	19.93	20.24	20.56	20.87	21.19
27	17.15	17.47	17.80	18.12	18.44	18.76	19.09	19.41	19.74	20.06	20.39	20.71	21.04	21.37	21.69	22.02
28	17.80	18.13	18.47	18.80	19.14	19.47	19.81	20.15	20.48	20.82	21.16	21.50	21.84	22.18	22.52	22.86
29	18.45	18.79	19.14	19.49	19.83	20.18	20.53	20.88	21.23	21.58	21.94	22.29	22.64	22.99	23.35	23.70
30	19.10	19.45	19.81	20.17	20.54	20.90	21.26	21.62	21.99	22.35	22.72	23.09	23.45	23.81	24.18	24.55
31	19.75	20.12	20.49	20.87	21.24	21.61	21.99	22.37	22.74	23.12	23.50	23.88	24.26	24.64	25.02	25.40
32	20.40	20.79	21.17	21.56	21.95	22.33	22.72	23.11	23.50	23.89	24.28	24.68	25.07	25.46	25.86	26.25
33	21.06	21.46	21.85	22.25	22.65	23.06	23.46	23.86	24.26	24.67	25.07	25.48	25.88	26.29	26.70	27.11
34	21.72	22.13	22.54	22.95	23.37	23.78	24.19	24.61	25.03	25.44	25.86	26.28	26.70	27.12	27.54	27.97
35	22.38	22.80	23.23	23.65	24.08	24.51	24.94	25.36	25.79	26.23	26.66	27.09	27.52	27.96	28.39	28.83
36	23.04	23.48	23.92	24.35	24.80	25.24	25.68	26.12	26.57	27.01	27.46	27.90	28.35	28.80	29.25	29.70
37	23.70	24.16	24.61	25.06	25.51	25.97	26.42	26.88	27.34	27.80	28.26	28.72	29.18	29.64	30.10	30.57
38	24.37	24.84	25.30	25.77	26.24	26.70	27.17	27.64	28.11	28.59	29.06	29.53	30.01	30.49	30.96	31.44
39	25.04	25.52	26.00	26.48	26.96	27.44	27.92	28.41	28.89	29.38	29.87	30.36	30.85	31.34	31.83	32.32
40	25.71	26.20	26.70	27.19	27.69	28.18	28.68	29.18	29.68	30.18	30.68	31.18	31.68	32.19	32.69	33.20
41	26.39	26.89	27.40	27.91	28.41	28.92	29.44	29.95	30.46	30.97	31.49	32.01	32.52	33.04	33.56	34.08
42	27.06	27.58	28.10	28.62	29.15	29.67	30.19	30.72	31.25	31.78	32.31	32.84	33.37	33.90	34.44	34.97
43	27.74	28.27	28.81	29.34	29.88	30.42	30.96	31.50	32.04	32.58	33.13	33.67	34.22	34.76	35.31	35.86
44	28.42	28.97	29.52	30.07	30.62	31.17	31.72	32.28	32.83	33.39	33.95	34.51	35.07	35.63	36.19	36.76
45	29.11	29.67	30.23	30.79	31.36	31.92	32.49	33.06	33.63	34.20	34.77	35.35	35.92	36.50	37.08	37.66
46	29.79	30.36	30.94	31.52	32.10	32.68	33.26	33.84	34.43	35.01	35.60	36.19	36.78	37.37	37.96	38.56
47	30.48	31.07	31.66	32.25	32.84	33.44	34.03	34.63	35.23	35.83	36.43	37.04	37.64	38.25	38.86	39.46
48	31.17	31.77	32.37	32.98	33.59	34.20	34.81	35.42	36.03	36.65	37.27	37.88	38.50	39.13	39.75	40.37
49	31.86	32.48	33.09	33.71	34.34	34.96	35.59	36.21	36.84	37.47	38.10	38.74	39.37	40.01	40.65	41.29
50	32.55	33.18	33.82	34.45	35.09	35.73	36.37	37.01	37.65	38.30	38.94	39.59	40.24	40.89	41.55	42.20
51	33.25	33.89	34.54	35.19	35.84	36.49	37.15	37.81	38.46	39.12	39.79	40.45	41.11	41.78	42.45	43.12
52	33.95	34.61	35.27	35.93	36.60	37.27	37.94	38.61	39.28	39.96	40.63	41.31	41.99	42.67	43.36	44.04
53	34.65	35.32	36.00	36.68	37.36	38.04	38.72	39.41	40.10	40.79	41.48	42.17	42.87	43.57	44.27	44.97
54	35.35	36.04	36.73	37.42	38.12	38.82	39.52	40.22	40.92	41.63	42.33	43.04	43.75	44.47	45.18	45.90
55	36.05	36.76	37.46	38.17	38.88	39.60	40.31	41.03	41.74	42.47	43.19	43.91	44.64	45.37	46.10	46.83
56	36.76	37.48	38.20	38.92	39.65	40.38	41.11	41.84	42.57	43.31	44.05	44.79	45.53	46.27	47.02	47.77
57	37.47	38.20	38.94	39.68	40.42	41.16	41.91	42.65	43.40	44.15	44.91	45.66	46.42	47.18	47.94	48.71
58	38.18	38.93	39.68	40.43	41.19	41.95	42.71	43.47	44.23	45.00	45.77	46.54	47.32	48.09	48.87	49.65
59	38.89	39.66	40.42	41.19	41.96	42.74	43.51	44.29	45.07	45.85	46.64	47.42	48.21	49.01	49.80	50.60
60	39.61	40.39	41.17	41.95	42.74	43.53	44.32	45.11	45.91	46.71	47.51	48.31	49.12	49.92	50.73	51.55

Monthly Payments
(Necessary to amortize a loan of $1,000)

Interest Rate	5 Years	10 Years	15 Years	20 Years	25 Years	30 Years	35 Years	40 Years
5 %	18.88	10.61	7.91	6.60	5.85	5.37	5.05	4.83
5¼	18.99	10.73	8.04	6.74	6.00	5.53	5.21	4.99
5½	19.11	10.86	8.18	6.88	6.15	5.68	5.38	5.16
5¾	19.22	10.98	8.31	7.03	6.30	5.84	5.54	5.33
6	19.34	11.11	8.44	7.17	6.45	6.00	5.71	5.51
6¼	19.45	11.23	8.58	7.31	6.60	6.16	5.88	5.68
6½	19.57	11.36	8.72	7.46	6.76	6.33	6.05	5.86
6¾	19.69	11.49	8.85	7.61	6.91	6.49	6.22	6.04
7	19.81	11.62	8.99	7.76	7.07	6.66	6.39	6.22
7¼	19.92	11.75	9.13	7.91	7.23	6.83	6.57	6.40
7½	20.04	11.88	9.28	8.06	7.39	7.00	6.75	6.59
7¾	20.16	12.01	9.42	8.21	7.56	7.17	6.93	6.77
8	20.28	12.14	9.56	8.37	7.72	7.34	7.11	6.96
8¼	20.40	12.27	9.71	8.53	7.89	7.52	7.29	7.15
8½	20.52	12.40	9.85	8.68	8.06	7.69	7.47	7.34
8¾	20.64	12.54	10.00	8.84	8.23	7.87	7.66	7.53
9	20.76	12.67	10.15	9.00	8.40	8.05	7.84	7.72
9¼	20.88	12.81	10.30	9.16	8.57	8.23	8.03	7.91
9½	21.01	12.94	10.45	9.33	8.74	8.41	8.22	8.11
9¾	21.13	13.08	10.60	9.49	8.92	8.60	8.41	8.30
10	21.25	13.22	10.75	9.66	9.09	8.78	8.60	8.50
10¼	21.38	13.36	10.90	9.82	9.27	8.97	8.79	8.69
10½	21.50	13.50	11.06	9.99	9.45	9.15	8.99	8.89
10¾	21.62	13.64	11.21	10.16	9.63	9.34	9.18	9.09
11	21.75	13.78	11.37	10.33	9.81	9.53	9.37	9.29
11¼	21.87	13.92	11.53	10.50	9.99	9.72	9.57	9.49
11½	22.00	14.06	11.69	10.67	10.17	9.91	9.77	9.69
11¾	22.12	14.21	11.85	10.84	10.35	10.10	9.96	9.89
12	22.25	14.35	12.01	11.02	10.54	10.29	10.16	10.09
12¼	22.38	14.50	12.17	11.19	10.72	10.48	10.36	10.29
12½	22.50	14.64	12.33	11.37	10.91	10.68	10.56	10.49
12¾	22.63	14.79	12.49	11.54	11.10	10.87	10.76	10.70
13	22.76	14.94	12.66	11.72	11.28	11.07	10.96	10.90
13¼	22.89	15.08	12.82	11.90	11.47	11.26	11.16	11.10
13½	23.01	15.23	12.99	12.08	11.66	11.46	11.36	11.31
13¾	23.14	15.38	13.15	12.26	11.85	11.66	11.56	11.51
14	23.27	15.53	13.32	12.44	12.04	11.85	11.76	11.72

TABLE 14-1

AMORTIZATION TABLE OF MORTGAGE PRINCIPAL AND INTEREST PER $1,000

Lending Ratio Guidelines

Mortgage Type	Housing Expense Ratio	Total Obligations Ratio
FHA	29%	41%
Conventional	28%	36%

Hypothetical Enterprises, Inc.
Balance Sheet
December 31, 199X

Assets		
Current Assets		
Cash	$ 13,000	
Accounts Receivable	32,500	
Merchandise Inventory	50,600	
Prepaid Expenses	1,200	
Supplies	4,000	
Total Current Assets		$101,300
Property, Plant, and Equipment		
Land	40,000	
Buildings	125,000	
Machinery and Equipment	60,000	
Total Property, Plant, and Equipment		225,000
Investments and Other Assets		
Investments	10,000	
Intangible Assets	5,000	
Total Investments and Other Assets		15,000
Total Assets		$341,300

Liabilities and Owner's Equity		
Current Liabilities		
Accounts Payable	$ 17,500	
Salaries Payable	5,400	
Taxes Payable	6,500	
Total Current Liabilities		$ 29,400
Long-Term Liabilities		
Mortgage Payable	115,000	
Debenture Bond	20,000	
Total Long-Term Liabilities		135,000
Total Liabilities		164,400
Stockholder's Equity		
Capital Stock	126,900	
Retained Earnings	50,000	
Total Stockholder's Equity		176,900
Total Liabilities and Stockholder's Equity		$341,300

Hypothetical Enterprises, Inc.
Income Statement
For the year ended December 31, 199X

Revenue		
Gross sales	$923,444	
Less: Sales Returns and Allowances	22,875	
Sales Discounts	3,625	
Net Sales		$896,944
Cost of Goods Sold		
Merchandise Inventory, Jan. 1	220,350	
Net Purchases	337,400	
Freight In	12,350	
Goods Available for Sale	570,100	
Less: Merchandise Inventory, Dec. 31	88,560	
Cost of Goods Sold		481,540
Gross Margin		415,404
Operating Expenses		
Salaries and Benefits	152,600	
Rent and Utilities	35,778	
Advertising and Promotion	32,871	
Insurance	8,258	
General and Administrative Expenses	41,340	
Depreciation	19,890	
Miscellaneous Expenses	14,790	
Total Operating Expenses		305,527
Income before Taxes		109,877
Income Tax		18,609
Net Income		$ 91,268

FIFO

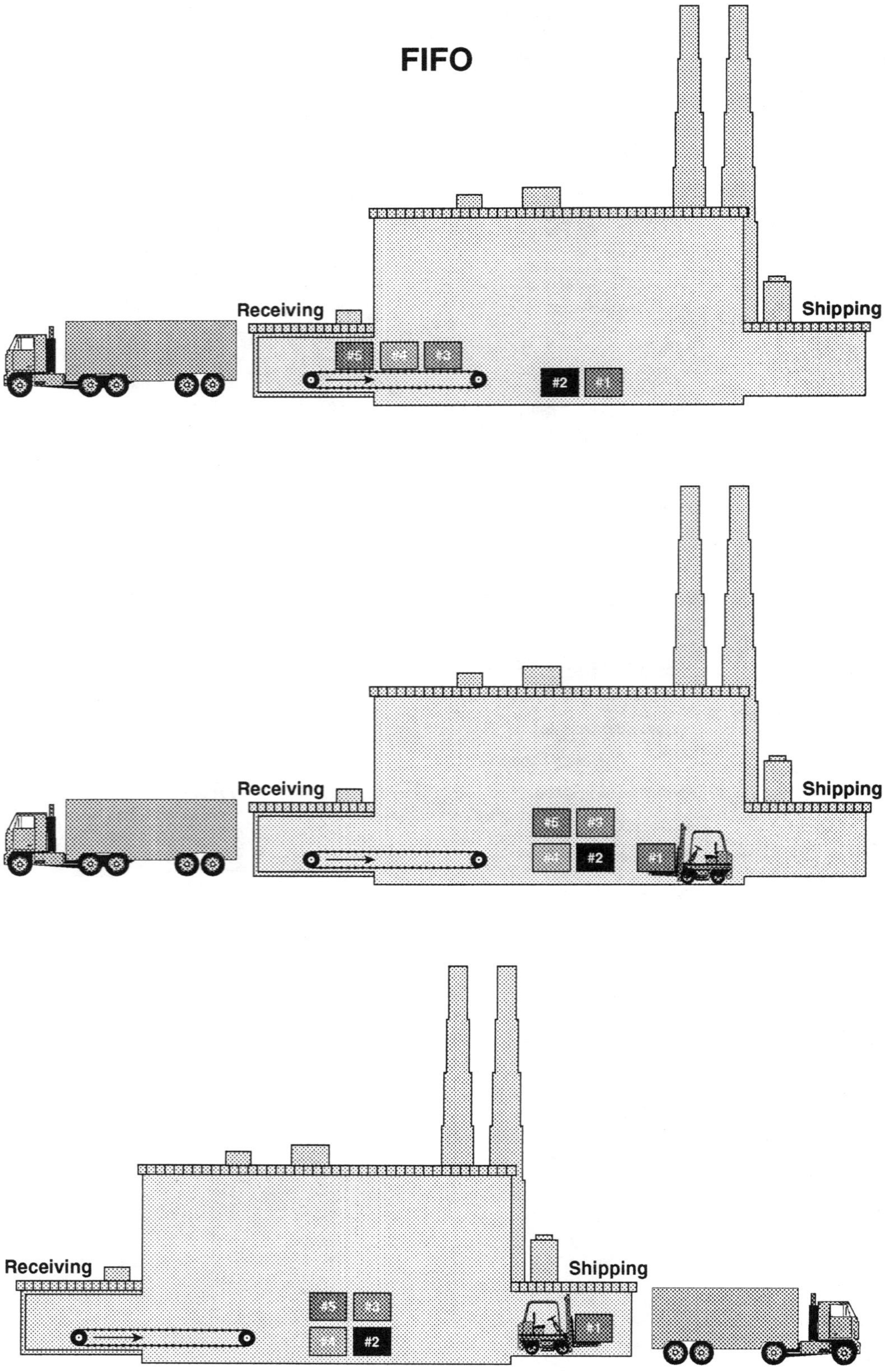

SOURCE: Adapted from Michael D. Tuttle, *Practical Business Math,* Sixth Edition (Dubuque, Iowa: William C. Brown, 1994), p. 293.

LIFO

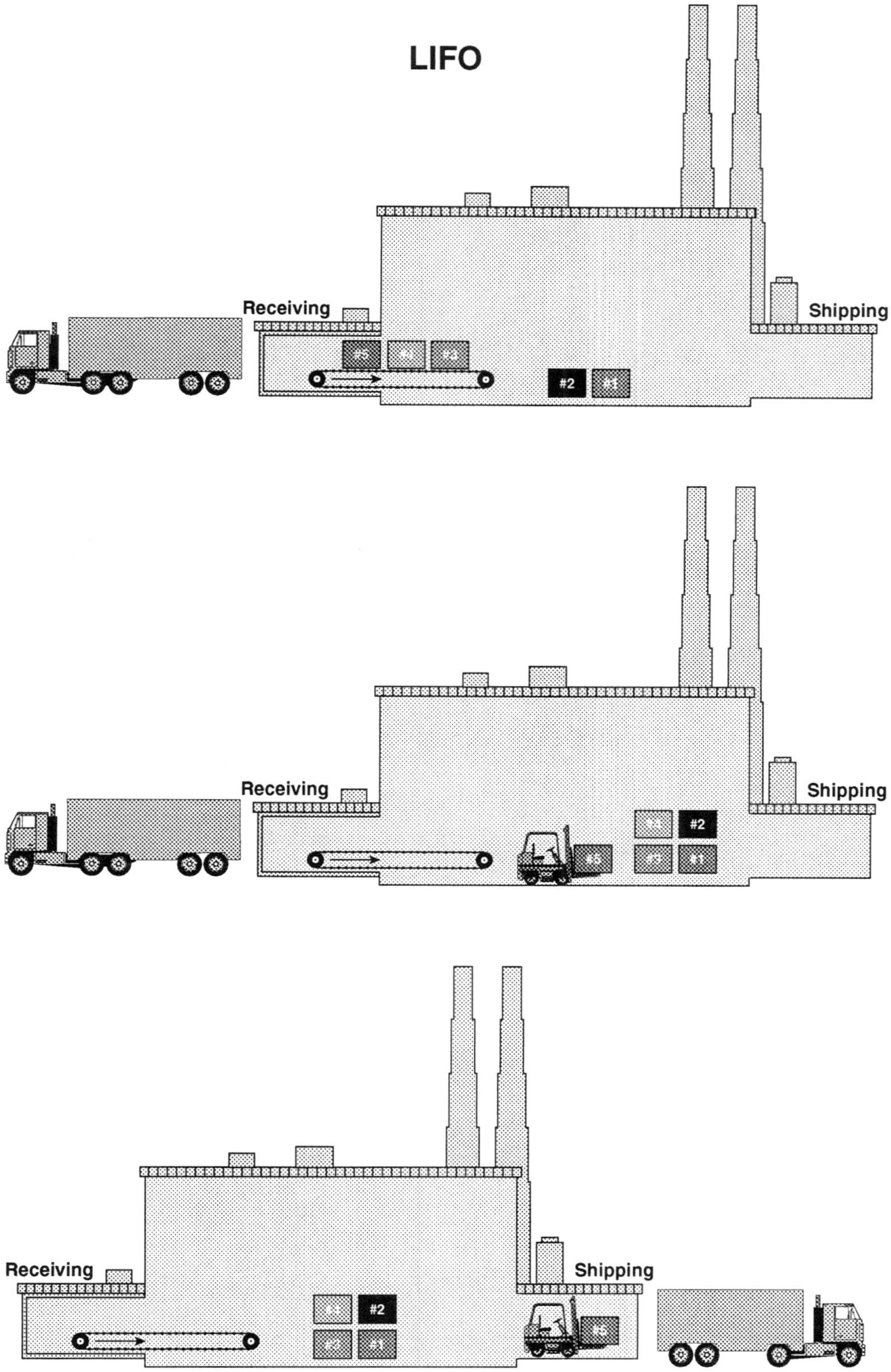
Receiving
Shipping
#2
Receiving
Shipping
#2
Receiving
Shipping
#2

3-Year Property	5-Year Property	7-Year Property
Over-the-road tractors Some horses and hogs Special handling devices for the manufacture of food and beverages Specialty tools used in the manufacture of motor vehicles Specialty tools used in the manufacture of finished products made of plastic, rubber, glass, and metal	Automobiles and taxis Buses and trucks Computers and peripherals Office machinery Research and experimental equipment Breeding or dairy cattle Sheep and goats Airplanes (except those in commercial use) Drilling and timber cutting equipment Construction equipment	Office furniture and fixtures Railroad cars and engines Commercial airplanes Equipment used in mining, petroleum drilling, and natural gas exploration Equipment used in the manufacture of wood, pulp, and paper products Equipment used to manufacture aerospace products

10-Year Property	15-Year Property	20-Year Property
Vessels, barges, and tugs Single-purpose agricultural structures Trees and vines bearing fruits or nuts Equipment for grain, sugar, and vegetable oil products	Depreciable improvements made to land such as shrubbery, fences, roads, and bridges Equipment used to manufacture cement Gas utility pipelines	Farm buildings Railroad structures and improvements Communication cable and long-line systems Water utility plants and equipment

31.5-Year Property	39-Year Property
Placed into service before May 13, 1993: Nonresidential real estate Office in the home	*Placed into service after May 12, 1993:* Nonresidential real estate Office in the home

TABLE 17-1

MACRS PROPERTY CLASSES
GENERAL DEPRECIATION SYSTEM

Recovery Year	Depreciation Rate for Property Class					
	3-year	5-year	7-year	10-year	15-year	20-year
1	33.33%	20.00%	14.29%	10.00%	5.00%	3.750%
2	44.45	32.00	24.49	18.00	9.50	7.219
3	14.81	19.20	17.49	14.40	8.55	6.677
4	7.41	11.52	12.49	11.52	7.70	6.177
5		11.52	8.93	9.22	6.93	5.713
6		5.76	8.92	7.37	6.23	5.285
7			8.93	6.55	5.90	4.888
8			4.46	6.55	5.90	4.522
9				6.56	5.91	4.462
10				6.55	5.90	4.461
11				3.28	5.91	4.462
12					5.90	4.461
13					5.91	4.462
14					5.90	4.461
15					5.91	4.462
16					2.95	4.461
17						4.462
18						4.461
19						4.462
20						4.461
21						2.231

TABLE 17-2

COST RECOVERY
PERCENTAGE TABLE
MACRS

SALES TAX BRACKETS
ON ALL 6 1/2% TAXABLE TRANSACTIONS
(FLORIDA DEPT. OF REVENUE)

Amount of Sale	Tax	Amount of Sale	Tax
.10- .15	.01	5.08- 5.23	.34
.16- .30	.02	5.24- 5.38	.35
.31- .46	.03	5.39- 5.53	.36
.47- .61	.04	5.54- 5.69	.37
.62- .76	.05	5.70- 5.84	.38
.77- .92	.06	5.85- 6.09	.39
.93- 1.07	.07	6.10- 6.15	.40
1.08- 1.23	.08	6.16- 6.30	.41
1.24- 1.38	.09	6.31- 6.46	.42
1.39- 1.53	.10	6.47- 6.61	.43
1.54- 1.69	.11	6.62- 6.76	.44
1.70- 1.84	.12	6.77- 6.92	.45
1.85- 2.09	.13	6.93- 7.07	.46
2.10- 2.15	.14	7.08- 7.23	.47
2.16- 2.30	.15	7.24- 7.38	.48
2.31- 2.46	.16	7.39- 7.53	.49
2.47- 2.61	.17	7.54- 7.69	.50
2.62- 2.76	.18	7.70- 7.84	.51
2.77- 2.92	.19	7.85- 8.09	.52
2.93- 3.07	.20	8.10- 8.15	.53
3.08- 3.23	.21	8.16- 8.30	.54
3.24- 3.38	.22	8.31- 8.46	.55
3.39- 3.53	.23	8.47- 8.61	.56
3.54- 3.69	.24	8.62- 8.76	.57
3.70- 3.84	.25	8.77- 8.92	.58
3.85- 4.09	.26	8.93- 9.07	.59
4.10- 4.15	.27	9.08- 9.23	.60
4.16- 4.30	.28	9.24- 9.38	.61
4.31- 4.46	.29	9.39- 9.53	.62
4.47- 4.61	.30	9.54- 9.69	.63
4.62- 4.76	.31	9.70- 9.84	.64
4.77- 4.92	.32	9.85- 10.09	.65
4.93- 5.07	.33		

SALES TAX BRACKETS
ON ALL 6 1/2% TAXABLE TRANSACTIONS
(FLORIDA DEPT. OF REVENUE)

Amount of Sale	Tax	Amount of Sale	Tax
10.10- 10.15	.66	15.08- 15.23	.99
10.16- 10.30	.67	15.24- 15.38	1.00
10.31- 10.46	.68	15.39- 15.53	1.01
10.47- 10.61	.69	15.54- 15.69	1.02
10.62- 10.76	.70	15.70- 15.84	1.03
10.77- 10.92	.71	15.85- 16.09	1.04
10.93- 11.07	.72	16.10- 16.15	1.05
11.08- 11.23	.73	16.16- 16.30	1.06
11.24- 11.38	.74	16.31- 16.46	1.07
11.39- 11.53	.75	16.47- 16.61	1.08
11.54- 11.69	.76	16.62- 16.76	1.09
11.70- 11.84	.77	16.77- 16.92	1.10
11.85- 12.09	.78	16.93- 17.07	1.11
12.10- 12.15	.79	17.08- 17.23	1.12
12.16- 12.30	.80	17.24- 17.38	1.13
12.31- 12.46	.81	17.39- 17.53	1.14
12.47- 12.61	.82	17.54- 17.69	1.15
12.62- 12.76	.83	17.70- 17.84	1.16
12.77- 12.92	.84	17.85- 18.09	1.17
12.93 13.07	.85	18.10- 18.15	1.18
13.08- 13.23	.86	18.16- 18.30	1.19
13.24- 13.38	.87	18.31- 18.46	1.20
13.39- 13.53	.88	18.47- 18.61	1.21
13.54- 13.69	.89	18.62- 18.76	1.22
13.70- 13.84	.90	18.77- 18.92	1.23
13.85- 14.09	.91	18.93- 19.07	1.24
14.10- 14.15	.92	19.08- 19.23	1.25
14.16- 14.30	.93	19.24- 19.38	1.26
14.31- 14.46	.94	19.39- 19.53	1.27
14.47- 14.61	.95	19.54- 19.69	1.28
14.62- 14.76	.96	19.70- 19.84	1.29
14.77- 14.92	.97	19.85- 20.09	1.30
14.93- 15.07	.98		

EXHIBIT 18-1

SALES TAX BRACKETS

1995 Standard Deductions:

Single	$3,900
Married filing jointly or Qualifying widow(er)	6,550
Married, filing separately	3,275
Head of household	5,750
65 or older, and/or blind	See IRS instructions to find standard deduction

Income	Wages, salaries, bonuses, commissions, tips, gratuities Interest and dividend income Rents, royalties, partnerships, S corporations, trusts Pensions and annuities Business income (or loss) Capital gain (or loss) from the sale or exchange of property Farm income Unemployment compensation, social security benefits Contest prizes, gambling winnings

Less

Adjustments to Income	Alimony payments Retirement fund payments—IRA, Keogh, 401K One-half of self-employment tax Self-employment health insurance Penalty on early withdrawal of savings

Equals

Adjusted Gross Income	Used in determining limits on certain itemized deductions, such as medical, dental, and employee expenses

Less

Deductions: Standard or Itemized	Medical and dental expenses (above 7.5% of adjusted gross income) Taxes paid: state and local income taxes; real estate taxes Home mortgage interest and points Charitable contributions Casualty and theft losses Moving expenses Unreimbursed employee expenses—union dues, job travel, education (above 2.0% of adjusted gross income)

and

Exemptions	Personal exemptions Dependents' exemptions

Equals

Taxable Income	Income on which the amount of income tax due is based. Used for Tax Table look-up and Tax Rate Schedule computation

EXHIBIT 18-2

PROCEDURE TO CALCULATE TAXABLE INCOME

1995 Tax Table—*Continued*

If line 37 (taxable income) is— At least	But less than	And you are— Single	Married filing jointly *	Married filing separately	Head of a household
		Your tax is—			
23,000					
23,000	**23,050**	3,454	3,454	3,912	3,454
23,050	**23,100**	3,461	3,461	3,926	3,461
23,100	**23,150**	3,469	3,469	3,940	3,469
23,150	**23,200**	3,476	3,476	3,954	3,476
23,200	**23,250**	3,484	3,484	3,968	3,484
23,250	**23,300**	3,491	3,491	3,982	3,491
23,300	**23,350**	3,499	3,499	3,996	3,499
23,350	**23,400**	3,510	3,506	4,010	3,506
23,400	**23,450**	3,524	3,514	4,024	3,514
23,450	**23,500**	3,538	3,521	4,038	3,521
23,500	**23,550**	3,552	3,529	4,052	3,529
23,550	**23,600**	3,566	3,536	4,066	3,536
23,600	**23,650**	3,580	3,544	4,080	3,544
23,650	**23,700**	3,594	3,551	4,094	3,551
23,700	**23,750**	3,608	3,559	4,108	3,559
23,750	**23,800**	3,622	3,566	4,122	3,566
23,800	**23,850**	3,636	3,574	4,136	3,574
23,850	**23,900**	3,650	3,581	4,150	3,581
23,900	**23,950**	3,664	3,589	4,164	3,589
23,950	**24,000**	3,678	3,596	4,178	3,596
24,000					
24,000	**24,050**	3,692	3,604	4,192	3,604
24,050	**24,100**	3,706	3,611	4,206	3,611
24,100	**24,150**	3,720	3,619	4,220	3,619
24,150	**24,200**	3,734	3,626	4,234	3,626
24,200	**24,250**	3,748	3,634	4,248	3,634
24,250	**24,300**	3,762	3,641	4,262	3,641
24,300	**24,350**	3,776	3,649	4,276	3,649
24,350	**24,400**	3,790	3,656	4,290	3,656
24,400	**24,450**	3,804	3,664	4,304	3,664
24,450	**24,500**	3,818	3,671	4,318	3,671
24,500	**24,550**	3,832	3,679	4,332	3,679
24,550	**24,600**	3,846	3,686	4,346	3,686
24,600	**24,650**	3,860	3,694	4,360	3,694
24,650	**24,700**	3,874	3,701	4,374	3,701
24,700	**24,750**	3,888	3,709	4,388	3,709
24,750	**24,800**	3,902	3,716	4,402	3,716
24,800	**24,850**	3,916	3,724	4,416	3,724
24,850	**24,900**	3,930	3,731	4,430	3,731
24,900	**24,950**	3,944	3,739	4,444	3,739
24,950	**25,000**	3,958	3,746	4,458	3,746
25,000					
25,000	**25,050**	3,972	3,754	4,472	3,754
25,050	**25,100**	3,986	3,761	4,486	3,761
25,100	**25,150**	4,000	3,769	4,500	3,769
25,150	**25,200**	4,014	3,776	4,514	3,776
25,200	**25,250**	4,028	3,784	4,528	3,784
25,250	**25,300**	4,042	3,791	4,542	3,791
25,300	**25,350**	4,056	3,799	4,556	3,799
25,350	**25,400**	4,070	3,806	4,570	3,806
25,400	**25,450**	4,084	3,814	4,584	3,814
25,450	**25,500**	4,098	3,821	4,598	3,821
25,500	**25,550**	4,112	3,829	4,612	3,829
25,550	**25,600**	4,126	3,836	4,626	3,836
25,600	**25,650**	4,140	3,844	4,640	3,844
25,650	**25,700**	4,154	3,851	4,654	3,851
25,700	**25,750**	4,168	3,859	4,668	3,859
25,750	**25,800**	4,182	3,866	4,682	3,866
25,800	**25,850**	4,196	3,874	4,696	3,874
25,850	**25,900**	4,210	3,881	4,710	3,881
25,900	**25,950**	4,224	3,889	4,724	3,889
25,950	**26,000**	4,238	3,896	4,738	3,896
26,000					
26,000	**26,050**	4,252	3,904	4,752	3,904
26,050	**26,100**	4,266	3,911	4,766	3,911
26,100	**26,150**	4,280	3,919	4,780	3,919
26,150	**26,200**	4,294	3,926	4,794	3,926
26,200	**26,250**	4,308	3,934	4,808	3,934
26,250	**26,300**	4,322	3,941	4,822	3,941
26,300	**26,350**	4,336	3,949	4,836	3,949
26,350	**26,400**	4,350	3,956	4,850	3,956
26,400	**26,450**	4,364	3,964	4,864	3,964
26,450	**26,500**	4,378	3,971	4,878	3,971
26,500	**26,550**	4,392	3,979	4,892	3,979
26,550	**26,600**	4,406	3,986	4,906	3,986
26,600	**26,650**	4,420	3,994	4,920	3,994
26,650	**26,700**	4,434	4,001	4,934	4,001
26,700	**26,750**	4,448	4,009	4,948	4,009
26,750	**26,800**	4,462	4,016	4,962	4,016
26,800	**26,850**	4,476	4,024	4,976	4,024
26,850	**26,900**	4,490	4,031	4,990	4,031
26,900	**26,950**	4,504	4,039	5,004	4,039
26,950	**27,000**	4,518	4,046	5,018	4,046
27,000					
27,000	**27,050**	4,532	4,054	5,032	4,054
27,050	**27,100**	4,546	4,061	5,046	4,061
27,100	**27,150**	4,560	4,069	5,060	4,069
27,150	**27,200**	4,574	4,076	5,074	4,076
27,200	**27,250**	4,588	4,084	5,088	4,084
27,250	**27,300**	4,602	4,091	5,102	4,091
27,300	**27,350**	4,616	4,099	5,116	4,099
27,350	**27,400**	4,630	4,106	5,130	4,106
27,400	**27,450**	4,644	4,114	5,144	4,114
27,450	**27,500**	4,658	4,121	5,158	4,121
27,500	**27,550**	4,672	4,129	5,172	4,129
27,550	**27,600**	4,686	4,136	5,186	4,136
27,600	**27,650**	4,700	4,144	5,200	4,144
27,650	**27,700**	4,714	4,151	5,214	4,151
27,700	**27,750**	4,728	4,159	5,228	4,159
27,750	**27,800**	4,742	4,166	5,242	4,166
27,800	**27,850**	4,756	4,174	5,256	4,174
27,850	**27,900**	4,770	4,181	5,270	4,181
27,900	**27,950**	4,784	4,189	5,284	4,189
27,950	**28,000**	4,798	4,196	5,298	4,196
28,000					
28,000	**28,050**	4,812	4,204	5,312	4,204
28,050	**28,100**	4,826	4,211	5,326	4,211
28,100	**28,150**	4,840	4,219	5,340	4,219
28,150	**28,200**	4,854	4,226	5,354	4,226
28,200	**28,250**	4,868	4,234	5,368	4,234
28,250	**28,300**	4,882	4,241	5,382	4,241
28,300	**28,350**	4,896	4,249	5,396	4,249
28,350	**28,400**	4,910	4,256	5,410	4,256
28,400	**28,450**	4,924	4,264	5,424	4,264
28,450	**28,500**	4,938	4,271	5,438	4,271
28,500	**28,550**	4,952	4,279	5,452	4,279
28,550	**28,600**	4,966	4,286	5,466	4,286
28,600	**28,650**	4,980	4,294	5,480	4,294
28,650	**28,700**	4,994	4,301	5,494	4,301
28,700	**28,750**	5,008	4,309	5,508	4,309
28,750	**28,800**	5,022	4,316	5,522	4,316
28,800	**28,850**	5,036	4,324	5,536	4,324
28,850	**28,900**	5,050	4,331	5,550	4,331
28,900	**28,950**	5,064	4,339	5,564	4,339
28,950	**29,000**	5,078	4,346	5,578	4,346
29,000					
29,000	**29,050**	5,092	4,354	5,592	4,354
29,050	**29,100**	5,106	4,361	5,606	4,361
29,100	**29,150**	5,120	4,369	5,620	4,369
29,150	**29,200**	5,134	4,376	5,634	4,376
29,200	**29,250**	5,148	4,384	5,648	4,384
29,250	**29,300**	5,162	4,391	5,662	4,391
29,300	**29,350**	5,176	4,399	5,676	4,399
29,350	**29,400**	5,190	4,406	5,690	4,406
29,400	**29,450**	5,204	4,414	5,704	4,414
29,450	**29,500**	5,218	4,421	5,718	4,421
29,500	**29,550**	5,232	4,429	5,732	4,429
29,550	**29,600**	5,246	4,436	5,746	4,436
29,600	**29,650**	5,260	4,444	5,760	4,444
29,650	**29,700**	5,274	4,451	5,774	4,451
29,700	**29,750**	5,288	4,459	5,788	4,459
29,750	**29,800**	5,302	4,466	5,802	4,466
29,800	**29,850**	5,316	4,474	5,816	4,474
29,850	**29,900**	5,330	4,481	5,830	4,481
29,900	**29,950**	5,344	4,489	5,844	4,489
29,950	**30,000**	5,358	4,496	5,858	4,496
30,000					
30,000	**30,050**	5,372	4,504	5,872	4,504
30,050	**30,100**	5,386	4,511	5,886	4,511
30,100	**30,150**	5,400	4,519	5,900	4,519
30,150	**30,200**	5,414	4,526	5,914	4,526
30,200	**30,250**	5,428	4,534	5,928	4,534
30,250	**30,300**	5,442	4,541	5,942	4,541
30,300	**30,350**	5,456	4,549	5,956	4,549
30,350	**30,400**	5,470	4,556	5,970	4,556
30,400	**30,450**	5,484	4,564	5,984	4,564
30,450	**30,500**	5,498	4,571	5,998	4,571
30,500	**30,550**	5,512	4,579	6,012	4,579
30,550	**30,600**	5,526	4,586	6,026	4,586
30,600	**30,650**	5,540	4,594	6,040	4,594
30,650	**30,700**	5,554	4,601	6,054	4,601
30,700	**30,750**	5,568	4,609	6,068	4,609
30,750	**30,800**	5,582	4,616	6,082	4,616
30,800	**30,850**	5,596	4,624	6,096	4,624
30,850	**30,900**	5,610	4,631	6,110	4,631
30,900	**30,950**	5,624	4,639	6,124	4,639
30,950	**31,000**	5,638	4,646	6,138	4,646
31,000					
31,000	**31,050**	5,652	4,654	6,152	4,654
31,050	**31,100**	5,666	4,661	6,166	4,661
31,100	**31,150**	5,680	4,669	6,180	4,669
31,150	**31,200**	5,694	4,676	6,194	4,676
31,200	**31,250**	5,708	4,684	6,208	4,684
31,250	**31,300**	5,722	4,691	6,222	4,695
31,300	**31,350**	5,736	4,699	6,236	4,709
31,350	**31,400**	5,750	4,706	6,250	4,723
31,400	**31,450**	5,764	4,714	6,264	4,737
31,450	**31,500**	5,778	4,721	6,278	4,751
31,500	**31,550**	5,792	4,729	6,292	4,765
31,550	**31,600**	5,806	4,736	6,306	4,779
31,600	**31,650**	5,820	4,744	6,320	4,793
31,650	**31,700**	5,834	4,751	6,334	4,807
31,700	**31,750**	5,848	4,759	6,348	4,821
31,750	**31,800**	5,862	4,766	6,362	4,835
31,800	**31,850**	5,876	4,774	6,376	4,849
31,850	**31,900**	5,890	4,781	6,390	4,863
31,900	**31,950**	5,904	4,789	6,404	4,877
31,950	**32,000**	5,918	4,796	6,418	4,891

* This column must also be used by a qualifying widow(er).

Continued on next page

EXHIBIT 18-3 TAX TABLE

1995 Tax Table—*Continued*

If line 37 (taxable income) is—		And you are—			
At least	But less than	Single	Married filing jointly *	Married filing sepa-rately	Head of a house-hold
		Your tax is—			
32,000					
32,000	**32,050**	5,932	4,804	6,432	4,905
32,050	**32,100**	5,946	4,811	6,446	4,919
32,100	**32,150**	5,960	4,819	6,460	4,933
32,150	**32,200**	5,974	4,826	6,474	4,947
32,200	**32,250**	5,988	4,834	6,488	4,961
32,250	**32,300**	6,002	4,841	6,502	4,975
32,300	**32,350**	6,016	4,849	6,516	4,989
32,350	**32,400**	6,030	4,856	6,530	5,003
32,400	**32,450**	6,044	4,864	6,544	5,017
32,450	**32,500**	6,058	4,871	6,558	5,031
32,500	**32,550**	6,072	4,879	6,572	5,045
32,550	**32,600**	6,086	4,886	6,586	5,059
32,600	**32,650**	6,100	4,894	6,600	5,073
32,650	**32,700**	6,114	4,901	6,614	5,087
32,700	**32,750**	6,128	4,909	6,628	5,101
32,750	**32,800**	6,142	4,916	6,642	5,115
32,800	**32,850**	6,156	4,924	6,656	5,129
32,850	**32,900**	6,170	4,931	6,670	5,143
32,900	**32,950**	6,184	4,939	6,684	5,157
32,950	**33,000**	6,198	4,946	6,698	5,171
33,000					
33,000	**33,050**	6,212	4,954	6,712	5,185
33,050	**33,100**	6,226	4,961	6,726	5,199
33,100	**33,150**	6,240	4,969	6,740	5,213
33,150	**33,200**	6,254	4,976	6,754	5,227
33,200	**33,250**	6,268	4,984	6,768	5,241
33,250	**33,300**	6,282	4,991	6,782	5,255
33,300	**33,350**	6,296	4,999	6,796	5,269
33,350	**33,400**	6,310	5,006	6,810	5,283
33,400	**33,450**	6,324	5,014	6,824	5,297
33,450	**33,500**	6,338	5,021	6,838	5,311
33,500	**33,550**	6,352	5,029	6,852	5,325
33,550	**33,600**	6,366	5,036	6,866	5,339
33,600	**33,650**	6,380	5,044	6,880	5,353
33,650	**33,700**	6,394	5,051	6,894	5,367
33,700	**33,750**	6,408	5,059	6,908	5,381
33,750	**33,800**	6,422	5,066	6,922	5,395
33,800	**33,850**	6,436	5,074	6,936	5,409
33,850	**33,900**	6,450	5,081	6,950	5,423
33,900	**33,950**	6,464	5,089	6,964	5,437
33,950	**34,000**	6,478	5,096	6,978	5,451
34,000					
34,000	**34,050**	6,492	5,104	6,992	5,465
34,050	**34,100**	6,506	5,111	7,006	5,479
34,100	**34,150**	6,520	5,119	7,020	5,493
34,150	**34,200**	6,534	5,126	7,034	5,507
34,200	**34,250**	6,548	5,134	7,048	5,521
34,250	**34,300**	6,562	5,141	7,062	5,535
34,300	**34,350**	6,576	5,149	7,076	5,549
34,350	**34,400**	6,590	5,156	7,090	5,563
34,400	**34,450**	6,604	5,164	7,104	5,577
34,450	**34,500**	6,618	5,171	7,118	5,591
34,500	**34,550**	6,632	5,179	7,132	5,605
34,550	**34,600**	6,646	5,186	7,146	5,619
34,600	**34,650**	6,660	5,194	7,160	5,633
34,650	**34,700**	6,674	5,201	7,174	5,647
34,700	**34,750**	6,688	5,209	7,188	5,661
34,750	**34,800**	6,702	5,216	7,202	5,675
34,800	**34,850**	6,716	5,224	7,216	5,689
34,850	**34,900**	6,730	5,231	7,230	5,703
34,900	**34,950**	6,744	5,239	7,244	5,717
34,950	**35,000**	6,758	5,246	7,258	5,731

If line 37 (taxable income) is—		And you are—			
At least	But less than	Single	Married filing jointly *	Married filing sepa-rately	Head of a house-hold
		Your tax is—			
35,000					
35,000	**35,050**	6,772	5,254	7,272	5,745
35,050	**35,100**	6,786	5,261	7,286	5,759
35,100	**35,150**	6,800	5,269	7,300	5,773
35,150	**35,200**	6,814	5,276	7,314	5,787
35,200	**35,250**	6,828	5,284	7,328	5,801
35,250	**35,300**	6,842	5,291	7,342	5,815
35,300	**35,350**	6,856	5,299	7,356	5,829
35,350	**35,400**	6,870	5,306	7,370	5,843
35,400	**35,450**	6,884	5,314	7,384	5,857
35,450	**35,500**	6,898	5,321	7,398	5,871
35,500	**35,550**	6,912	5,329	7,412	5,885
35,550	**35,600**	6,926	5,336	7,426	5,899
35,600	**35,650**	6,940	5,344	7,440	5,913
35,650	**35,700**	6,954	5,351	7,454	5,927
35,700	**35,750**	6,968	5,359	7,468	5,941
35,750	**35,800**	6,982	5,366	7,482	5,955
35,800	**35,850**	6,996	5,374	7,496	5,969
35,850	**35,900**	7,010	5,381	7,510	5,983
35,900	**35,950**	7,024	5,389	7,524	5,997
35,950	**36,000**	7,038	5,396	7,538	6,011
36,000					
36,000	**36,050**	7,052	5,404	7,552	6,025
36,050	**36,100**	7,066	5,411	7,566	6,039
36,100	**36,150**	7,080	5,419	7,580	6,053
36,150	**36,200**	7,094	5,426	7,594	6,067
36,200	**36,250**	7,108	5,434	7,608	6,081
36,250	**36,300**	7,122	5,441	7,622	6,095
36,300	**36,350**	7,136	5,449	7,636	6,109
36,350	**36,400**	7,150	5,456	7,650	6,123
36,400	**36,450**	7,164	5,464	7,664	6,137
36,450	**36,500**	7,178	5,471	7,678	6,151
36,500	**36,550**	7,192	5,479	7,692	6,165
36,550	**36,600**	7,206	5,486	7,706	6,179
36,600	**36,650**	7,220	5,494	7,720	6,193
36,650	**36,700**	7,234	5,501	7,734	6,207
36,700	**36,750**	7,248	5,509	7,748	6,221
36,750	**36,800**	7,262	5,516	7,762	6,235
36,800	**36,850**	7,276	5,524	7,776	6,249
36,850	**36,900**	7,290	5,531	7,790	6,263
36,900	**36,950**	7,304	5,539	7,804	6,277
36,950	**37,000**	7,318	5,546	7,818	6,291
37,000					
37,000	**37,050**	7,332	5,554	7,832	6,305
37,050	**37,100**	7,346	5,561	7,846	6,319
37,100	**37,150**	7,360	5,569	7,860	6,333
37,150	**37,200**	7,374	5,576	7,874	6,347
37,200	**37,250**	7,388	5,584	7,888	6,361
37,250	**37,300**	7,402	5,591	7,902	6,375
37,300	**37,350**	7,416	5,599	7,916	6,389
37,350	**37,400**	7,430	5,606	7,930	6,403
37,400	**37,450**	7,444	5,614	7,944	6,417
37,450	**37,500**	7,458	5,621	7,958	6,431
37,500	**37,550**	7,472	5,629	7,972	6,445
37,550	**37,600**	7,486	5,636	7,986	6,459
37,600	**37,650**	7,500	5,644	8,000	6,473
37,650	**37,700**	7,514	5,651	8,014	6,487
37,700	**37,750**	7,528	5,659	8,028	6,501
37,750	**37,800**	7,542	5,666	8,042	6,515
37,800	**37,850**	7,556	5,674	8,056	6,529
37,850	**37,900**	7,570	5,681	8,070	6,543
37,900	**37,950**	7,584	5,689	8,084	6,557
37,950	**38,000**	7,598	5,696	8,098	6,571

If line 37 (taxable income) is—		And you are—			
At least	But less than	Single	Married filing jointly *	Married filing sepa-rately	Head of a house-hold
		Your tax is—			
38,000					
38,000	**38,050**	7,612	5,704	8,112	6,585
38,050	**38,100**	7,626	5,711	8,126	6,599
38,100	**38,150**	7,640	5,719	8,140	6,613
38,150	**38,200**	7,654	5,726	8,154	6,627
38,200	**38,250**	7,668	5,734	8,168	6,641
38,250	**38,300**	7,682	5,741	8,182	6,655
38,300	**38,350**	7,696	5,749	8,196	6,669
38,350	**38,400**	7,710	5,756	8,210	6,683
38,400	**38,450**	7,724	5,764	8,224	6,697
38,450	**38,500**	7,738	5,771	8,238	6,711
38,500	**38,550**	7,752	5,779	8,252	6,725
38,550	**38,600**	7,766	5,786	8,266	6,739
38,600	**38,650**	7,780	5,794	8,280	6,753
38,650	**38,700**	7,794	5,801	8,294	6,767
38,700	**38,750**	7,808	5,809	8,308	6,781
38,750	**38,800**	7,822	5,816	8,322	6,795
38,800	**38,850**	7,836	5,824	8,336	6,809
38,850	**38,900**	7,850	5,831	8,350	6,823
38,900	**38,950**	7,864	5,839	8,364	6,837
38,950	**39,000**	7,878	5,846	8,378	6,851
39,000					
39,000	**39,050**	7,892	5,857	8,392	6,865
39,050	**39,100**	7,906	5,871	8,406	6,879
39,100	**39,150**	7,920	5,885	8,420	6,893
39,150	**39,200**	7,934	5,899	8,434	6,907
39,200	**39,250**	7,948	5,913	8,448	6,921
39,250	**39,300**	7,962	5,927	8,462	6,935
39,300	**39,350**	7,976	5,941	8,476	6,949
39,350	**39,400**	7,990	5,955	8,490	6,963
39,400	**39,450**	8,004	5,969	8,504	6,977
39,450	**39,500**	8,018	5,983	8,518	6,991
39,500	**39,550**	8,032	5,997	8,532	7,005
39,550	**39,600**	8,046	6,011	8,546	7,019
39,600	**39,650**	8,060	6,025	8,560	7,033
39,650	**39,700**	8,074	6,039	8,574	7,047
39,700	**39,750**	8,088	6,053	8,588	7,061
39,750	**39,800**	8,102	6,067	8,602	7,075
39,800	**39,850**	8,116	6,081	8,616	7,089
39,850	**39,900**	8,130	6,095	8,630	7,103
39,900	**39,950**	8,144	6,109	8,644	7,117
39,950	**40,000**	8,158	6,123	8,658	7,131
40,000					
40,000	**40,050**	8,172	6,137	8,672	7,145
40,050	**40,100**	8,186	6,151	8,686	7,159
40,100	**40,150**	8,200	6,165	8,700	7,173
40,150	**40,200**	8,214	6,179	8,714	7,187
40,200	**40,250**	8,228	6,193	8,728	7,201
40,250	**40,300**	8,242	6,207	8,742	7,215
40,300	**40,350**	8,256	6,221	8,756	7,229
40,350	**40,400**	8,270	6,235	8,770	7,243
40,400	**40,450**	8,284	6,249	8,784	7,257
40,450	**40,500**	8,298	6,263	8,798	7,271
40,500	**40,550**	8,312	6,277	8,812	7,285
40,550	**40,600**	8,326	6,291	8,826	7,299
40,600	**40,650**	8,340	6,305	8,840	7,313
40,650	**40,700**	8,354	6,319	8,854	7,327
40,700	**40,750**	8,368	6,333	8,868	7,341
40,750	**40,800**	8,382	6,347	8,882	7,355
40,800	**40,850**	8,396	6,361	8,896	7,369
40,850	**40,900**	8,410	6,375	8,910	7,383
40,900	**40,950**	8,424	6,389	8,924	7,397
40,950	**41,000**	8,438	6,403	8,938	7,411

* This column must also be used by a qualifying widow(er).

Continued on next page

EXHIBIT 18-3 (continued)

If line 37 (taxable income) is—		And you are—			
At least	But less than	Single	Married filing jointly *	Married filing sepa-rately	Head of a house-hold
		Your tax is—			
59,000					
59,000	59,050	13,566	11,457	14,349	12,465
59,050	59,100	13,581	11,471	14,365	12,479
59,100	59,150	13,597	11,485	14,380	12,493
59,150	59,200	13,612	11,499	14,396	12,507
59,200	59,250	13,628	11,513	14,411	12,521
59,250	59,300	13,643	11,527	14,427	12,535
59,300	59,350	13,659	11,541	14,442	12,549
59,350	59,400	13,674	11,555	14,458	12,563
59,400	59,450	13,690	11,569	14,473	12,577
59,450	59,500	13,705	11,583	14,489	12,591
59,500	59,550	13,721	11,597	14,504	12,605
59,550	59,600	13,736	11,611	14,520	12,619
59,600	59,650	13,752	11,625	14,535	12,633
59,650	59,700	13,767	11,639	14,551	12,647
59,700	59,750	13,783	11,653	14,566	12,661
59,750	59,800	13,798	11,667	14,582	12,675
59,800	59,850	13,814	11,681	14,597	12,689
59,850	59,900	13,829	11,695	14,613	12,703
59,900	59,950	13,845	11,709	14,628	12,717
59,950	60,000	13,860	11,723	14,644	12,731
60,000					
60,000	60,050	13,876	11,737	14,659	12,745
60,050	60,100	13,891	11,751	14,675	12,759
60,100	60,150	13,907	11,765	14,690	12,773
60,150	60,200	13,922	11,779	14,706	12,787
60,200	60,250	13,938	11,793	14,721	12,801
60,250	60,300	13,953	11,807	14,737	12,815
60,300	60,350	13,969	11,821	14,752	12,829
60,350	60,400	13,984	11,835	14,768	12,843
60,400	60,450	14,000	11,849	14,783	12,857
60,450	60,500	14,015	11,863	14,799	12,871
60,500	60,550	14,031	11,877	14,814	12,885
60,550	60,600	14,046	11,891	14,830	12,899
60,600	60,650	14,062	11,905	14,845	12,913
60,650	60,700	14,077	11,919	14,861	12,927
60,700	60,750	14,093	11,933	14,876	12,941
60,750	60,800	14,108	11,947	14,892	12,955
60,800	60,850	14,124	11,961	14,907	12,969
60,850	60,900	14,139	11,975	14,923	12,983
60,900	60,950	14,155	11,989	14,938	12,997
60,950	61,000	14,170	12,003	14,954	13,011
61,000					
61,000	61,050	14,186	12,017	14,969	13,025
61,050	61,100	14,201	12,031	14,985	13,039
61,100	61,150	14,217	12,045	15,000	13,053
61,150	61,200	14,232	12,059	15,016	13,067
61,200	61,250	14,248	12,073	15,031	13,081
61,250	61,300	14,263	12,087	15,047	13,095
61,300	61,350	14,279	12,101	15,062	13,109
61,350	61,400	14,294	12,115	15,078	13,123
61,400	61,450	14,310	12,129	15,093	13,137
61,450	61,500	14,325	12,143	15,109	13,151
61,500	61,550	14,341	12,157	15,124	13,165
61,550	61,600	14,356	12,171	15,140	13,179
61,600	61,650	14,372	12,185	15,155	13,193
61,650	61,700	14,387	12,199	15,171	13,207
61,700	61,750	14,403	12,213	15,186	13,221
61,750	61,800	14,418	12,227	15,202	13,235
61,800	61,850	14,434	12,241	15,217	13,249
61,850	61,900	14,449	12,255	15,233	13,263
61,900	61,950	14,465	12,269	15,248	13,277
61,950	62,000	14,480	12,283	15,264	13,291
62,000					
62,000	62,050	14,496	12,297	15,279	13,305
62,050	62,100	14,511	12,311	15,295	13,319
62,100	62,150	14,527	12,325	15,310	13,333
62,150	62,200	14,542	12,339	15,326	13,347
62,200	62,250	14,558	12,353	15,341	13,361
62,250	62,300	14,573	12,367	15,357	13,375
62,300	62,350	14,589	12,381	15,372	13,389
62,350	62,400	14,604	12,395	15,388	13,403
62,400	62,450	14,620	12,409	15,403	13,417
62,450	62,500	14,635	12,423	15,419	13,431
62,500	62,550	14,651	12,437	15,434	13,445
62,550	62,600	14,666	12,451	15,450	13,459
62,600	62,650	14,682	12,465	15,465	13,473
62,650	62,700	14,697	12,479	15,481	13,487
62,700	62,750	14,713	12,493	15,496	13,501
62,750	62,800	14,728	12,507	15,512	13,515
62,800	62,850	14,744	12,521	15,527	13,529
62,850	62,900	14,759	12,535	15,543	13,543
62,900	62,950	14,775	12,549	15,558	13,557
62,950	63,000	14,790	12,563	15,574	13,571
63,000					
63,000	63,050	14,806	12,577	15,589	13,585
63,050	63,100	14,821	12,591	15,605	13,599
63,100	63,150	14,837	12,605	15,620	13,613
63,150	63,200	14,852	12,619	15,636	13,627
63,200	63,250	14,868	12,633	15,651	13,641
63,250	63,300	14,883	12,647	15,667	13,655
63,300	63,350	14,899	12,661	15,682	13,669
63,350	63,400	14,914	12,675	15,698	13,683
63,400	63,450	14,930	12,689	15,713	13,697
63,450	63,500	14,945	12,703	15,729	13,711
63,500	63,550	14,961	12,717	15,744	13,725
63,550	63,600	14,976	12,731	15,760	13,739
63,600	63,650	14,992	12,745	15,775	13,753
63,650	63,700	15,007	12,759	15,791	13,767
63,700	63,750	15,023	12,773	15,806	13,781
63,750	63,800	15,038	12,787	15,822	13,795
63,800	63,850	15,054	12,801	15,837	13,809
63,850	63,900	15,069	12,815	15,853	13,823
63,900	63,950	15,085	12,829	15,868	13,837
63,950	64,000	15,100	12,843	15,884	13,851
64,000					
64,000	64,050	15,116	12,857	15,899	13,865
64,050	64,100	15,131	12,871	15,915	13,879
64,100	64,150	15,147	12,885	15,930	13,893
64,150	64,200	15,162	12,899	15,946	13,907
64,200	64,250	15,178	12,913	15,961	13,921
64,250	64,300	15,193	12,927	15,977	13,935
64,300	64,350	15,209	12,941	15,992	13,949
64,350	64,400	15,224	12,955	16,008	13,963
64,400	64,450	15,240	12,969	16,023	13,977
64,450	64,500	15,255	12,983	16,039	13,991
64,500	64,550	15,271	12,997	16,054	14,005
64,550	64,600	15,286	13,011	16,070	14,019
64,600	64,650	15,302	13,025	16,085	14,033
64,650	64,700	15,317	13,039	16,101	14,047
64,700	64,750	15,333	13,053	16,116	14,061
64,750	64,800	15,348	13,067	16,132	14,075
64,800	64,850	15,364	13,081	16,147	14,089
64,850	64,900	15,379	13,095	16,163	14,103
64,900	64,950	15,395	13,109	16,178	14,117
64,950	65,000	15,410	13,123	16,194	14,131
65,000					
65,000	65,050	15,426	13,137	16,209	14,145
65,050	65,100	15,441	13,151	16,225	14,159
65,100	65,150	15,457	13,165	16,240	14,173
65,150	65,200	15,472	13,179	16,256	14,187
65,200	65,250	15,488	13,193	16,271	14,201
65,250	65,300	15,503	13,207	16,287	14,215
65,300	65,350	15,519	13,221	16,302	14,229
65,350	65,400	15,534	13,235	16,318	14,243
65,400	65,450	15,550	13,249	16,333	14,257
65,450	65,500	15,565	13,263	16,349	14,271
65,500	65,550	15,581	13,277	16,364	14,285
65,550	65,600	15,596	13,291	16,380	14,299
65,600	65,650	15,612	13,305	16,395	14,313
65,650	65,700	15,627	13,319	16,411	14,327
65,700	65,750	15,643	13,333	16,426	14,341
65,750	65,800	15,658	13,347	16,442	14,355
65,800	65,850	15,674	13,361	16,457	14,369
65,850	65,900	15,689	13,375	16,473	14,383
65,900	65,950	15,705	13,389	16,488	14,397
65,950	66,000	15,720	13,403	16,504	14,411
66,000					
66,000	66,050	15,736	13,417	16,519	14,425
66,050	66,100	15,751	13,431	16,535	14,439
66,100	66,150	15,767	13,445	16,550	14,453
66,150	66,200	15,782	13,459	16,566	14,467
66,200	66,250	15,798	13,473	16,581	14,481
66,250	66,300	15,813	13,487	16,597	14,495
66,300	66,350	15,829	13,501	16,612	14,509
66,350	66,400	15,844	13,515	16,628	14,523
66,400	66,450	15,860	13,529	16,643	14,537
66,450	66,500	15,875	13,543	16,659	14,551
66,500	66,550	15,891	13,557	16,674	14,565
66,550	66,600	15,906	13,571	16,690	14,579
66,600	66,650	15,922	13,585	16,705	14,593
66,650	66,700	15,937	13,599	16,721	14,607
66,700	66,750	15,953	13,613	16,736	14,621
66,750	66,800	15,968	13,627	16,752	14,635
66,800	66,850	15,984	13,641	16,767	14,649
66,850	66,900	15,999	13,655	16,783	14,663
66,900	66,950	16,015	13,669	16,798	14,677
66,950	67,000	16,030	13,683	16,814	14,691
67,000					
67,000	67,050	16,046	13,697	16,829	14,705
67,050	67,100	16,061	13,711	16,845	14,719
67,100	67,150	16,077	13,725	16,860	14,733
67,150	67,200	16,092	13,739	16,876	14,747
67,200	67,250	16,108	13,753	16,891	14,761
67,250	67,300	16,123	13,767	16,907	14,775
67,300	67,350	16,139	13,781	16,922	14,789
67,350	67,400	16,154	13,795	16,938	14,803
67,400	67,450	16,170	13,809	16,953	14,817
67,450	67,500	16,185	13,823	16,969	14,831
67,500	67,550	16,201	13,837	16,984	14,845
67,550	67,600	16,216	13,851	17,000	14,859
67,600	67,650	16,232	13,865	17,015	14,873
67,650	67,700	16,247	13,879	17,031	14,887
67,700	67,750	16,263	13,893	17,046	14,901
67,750	67,800	16,278	13,907	17,062	14,915
67,800	67,850	16,294	13,921	17,077	14,929
67,850	67,900	16,309	13,935	17,093	14,943
67,900	67,950	16,325	13,949	17,108	14,957
67,950	68,000	16,340	13,963	17,124	14,971

* This column must also be used by a qualifying widow(er).

Continued on next page

If line 37 (taxable income) is—		And you are—			
At least	But less than	Single	Married filing jointly *	Married filing sepa-rately	Head of a house-hold
		Your tax is—			
68,000					
68,000	**68,050**	16,356	13,977	17,139	14,985
68,050	**68,100**	16,371	13,991	17,155	14,999
68,100	**68,150**	16,387	14,005	17,170	15,013
68,150	**68,200**	16,402	14,019	17,186	15,027
68,200	**68,250**	16,418	14,033	17,201	15,041
68,250	**68,300**	16,433	14,047	17,217	15,055
68,300	**68,350**	16,449	14,061	17,232	15,069
68,350	**68,400**	16,464	14,075	17,248	15,083
68,400	**68,450**	16,480	14,089	17,263	15,097
68,450	**68,500**	16,495	14,103	17,279	15,111
68,500	**68,550**	16,511	14,117	17,294	15,125
68,550	**68,600**	16,526	14,131	17,310	15,139
68,600	**68,650**	16,542	14,145	17,325	15,153
68,650	**68,700**	16,557	14,159	17,341	15,167
68,700	**68,750**	16,573	14,173	17,356	15,181
68,750	**68,800**	16,588	14,187	17,372	15,195
68,800	**68,850**	16,604	14,201	17,387	15,209
68,850	**68,900**	16,619	14,215	17,403	15,223
68,900	**68,950**	16,635	14,229	17,418	15,237
68,950	**69,000**	16,650	14,243	17,434	15,251
69,000					
69,000	**69,050**	16,666	14,257	17,449	15,265
69,050	**69,100**	16,681	14,271	17,465	15,279
69,100	**69,150**	16,697	14,285	17,480	15,293
69,150	**69,200**	16,712	14,299	17,496	15,307
69,200	**69,250**	16,728	14,313	17,511	15,321
69,250	**69,300**	16,743	14,327	17,527	15,335
69,300	**69,350**	16,759	14,341	17,542	15,349
69,350	**69,400**	16,774	14,355	17,558	15,363
69,400	**69,450**	16,790	14,369	17,573	15,377
69,450	**69,500**	16,805	14,383	17,589	15,391
69,500	**69,550**	16,821	14,397	17,604	15,405
69,550	**69,600**	16,836	14,411	17,620	15,419
69,600	**69,650**	16,852	14,425	17,635	15,433
69,650	**69,700**	16,867	14,439	17,651	15,447
69,700	**69,750**	16,883	14,453	17,666	15,461
69,750	**69,800**	16,898	14,467	17,682	15,475
69,800	**69,850**	16,914	14,481	17,697	15,489
69,850	**69,900**	16,929	14,495	17,713	15,503
69,900	**69,950**	16,945	14,509	17,728	15,517
69,950	**70,000**	16,960	14,523	17,744	15,531
70,000					
70,000	**70,050**	16,976	14,537	17,759	15,545
70,050	**70,100**	16,991	14,551	17,775	15,559
70,100	**70,150**	17,007	14,565	17,790	15,573
70,150	**70,200**	17,022	14,579	17,806	15,587
70,200	**70,250**	17,038	14,593	17,821	15,601
70,250	**70,300**	17,053	14,607	17,837	15,615
70,300	**70,350**	17,069	14,621	17,852	15,629
70,350	**70,400**	17,084	14,635	17,868	15,643
70,400	**70,450**	17,100	14,649	17,883	15,657
70,450	**70,500**	17,115	14,663	17,899	15,671
70,500	**70,550**	17,131	14,677	17,914	15,685
70,550	**70,600**	17,146	14,691	17,930	15,699
70,600	**70,650**	17,162	14,705	17,945	15,713
70,650	**70,700**	17,177	14,719	17,961	15,727
70,700	**70,750**	17,193	14,733	17,976	15,741
70,750	**70,800**	17,208	14,747	17,992	15,755
70,800	**70,850**	17,224	14,761	18,007	15,769
70,850	**70,900**	17,239	14,775	18,023	15,783
70,900	**70,950**	17,255	14,789	18,038	15,797
70,950	**71,000**	17,270	14,803	18,054	15,811

If line 37 (taxable income) is—		And you are—			
At least	But less than	Single	Married filing jointly *	Married filing sepa-rately	Head of a house-hold
		Your tax is—			
71,000					
71,000	**71,050**	17,286	14,817	18,069	15,825
71,050	**71,100**	17,301	14,831	18,085	15,839
71,100	**71,150**	17,317	14,845	18,100	15,853
71,150	**71,200**	17,332	14,859	18,116	15,867
71,200	**71,250**	17,348	14,873	18,131	15,881
71,250	**71,300**	17,363	14,887	18,147	15,895
71,300	**71,350**	17,379	14,901	18,162	15,909
71,350	**71,400**	17,394	14,915	18,178	15,923
71,400	**71,450**	17,410	14,929	18,193	15,937
71,450	**71,500**	17,425	14,943	18,209	15,951
71,500	**71,550**	17,441	14,957	18,224	15,965
71,550	**71,600**	17,456	14,971	18,240	15,979
71,600	**71,650**	17,472	14,985	18,255	15,993
71,650	**71,700**	17,487	14,999	18,271	16,007
71,700	**71,750**	17,503	15,013	18,286	16,021
71,750	**71,800**	17,518	15,027	18,302	16,035
71,800	**71,850**	17,534	15,041	18,318	16,049
71,850	**71,900**	17,549	15,055	18,336	16,063
71,900	**71,950**	17,565	15,069	18,354	16,077
71,950	**72,000**	17,580	15,083	18,372	16,091
72,000					
72,000	**72,050**	17,596	15,097	18,390	16,105
72,050	**72,100**	17,611	15,111	18,408	16,119
72,100	**72,150**	17,627	15,125	18,426	16,133
72,150	**72,200**	17,642	15,139	18,444	16,147
72,200	**72,250**	17,658	15,153	18,462	16,161
72,250	**72,300**	17,673	15,167	18,480	16,175
72,300	**72,350**	17,689	15,181	18,498	16,189
72,350	**72,400**	17,704	15,195	18,516	16,203
72,400	**72,450**	17,720	15,209	18,534	16,217
72,450	**72,500**	17,735	15,223	18,552	16,231
72,500	**72,550**	17,751	15,237	18,570	16,245
72,550	**72,600**	17,766	15,251	18,588	16,259
72,600	**72,650**	17,782	15,265	18,606	16,273
72,650	**72,700**	17,797	15,279	18,624	16,287
72,700	**72,750**	17,813	15,293	18,642	16,301
72,750	**72,800**	17,828	15,307	18,660	16,315
72,800	**72,850**	17,844	15,321	18,678	16,329
72,850	**72,900**	17,859	15,335	18,696	16,343
72,900	**72,950**	17,875	15,349	18,714	16,357
72,950	**73,000**	17,890	15,363	18,732	16,371
73,000					
73,000	**73,050**	17,906	15,377	18,750	16,385
73,050	**73,100**	17,921	15,391	18,768	16,399
73,100	**73,150**	17,937	15,405	18,786	16,413
73,150	**73,200**	17,952	15,419	18,804	16,427
73,200	**73,250**	17,968	15,433	18,822	16,441
73,250	**73,300**	17,983	15,447	18,840	16,455
73,300	**73,350**	17,999	15,461	18,858	16,469
73,350	**73,400**	18,014	15,475	18,876	16,483
73,400	**73,450**	18,030	15,489	18,894	16,497
73,450	**73,500**	18,045	15,503	18,912	16,511
73,500	**73,550**	18,061	15,517	18,930	16,525
73,550	**73,600**	18,076	15,531	18,948	16,539
73,600	**73,650**	18,092	15,545	18,966	16,553
73,650	**73,700**	18,107	15,559	18,984	16,567
73,700	**73,750**	18,123	15,573	19,002	16,581
73,750	**73,800**	18,138	15,587	19,020	16,595
73,800	**73,850**	18,154	15,601	19,038	16,609
73,850	**73,900**	18,169	15,615	19,056	16,623
73,900	**73,950**	18,185	15,629	19,074	16,637
73,950	**74,000**	18,200	15,643	19,092	16,651

If line 37 (taxable income) is—		And you are—			
At least	But less than	Single	Married filing jointly *	Married filing sepa-rately	Head of a house-hold
		Your tax is—			
74,000					
74,000	**74,050**	18,216	15,657	19,110	16,665
74,050	**74,100**	18,231	15,671	19,128	16,679
74,100	**74,150**	18,247	15,685	19,146	16,693
74,150	**74,200**	18,262	15,699	19,164	16,707
74,200	**74,250**	18,278	15,713	19,182	16,721
74,250	**74,300**	18,293	15,727	19,200	16,735
74,300	**74,350**	18,309	15,741	19,218	16,749
74,350	**74,400**	18,324	15,755	19,236	16,763
74,400	**74,450**	18,340	15,769	19,254	16,777
74,450	**74,500**	18,355	15,783	19,272	16,791
74,500	**74,550**	18,371	15,797	19,290	16,805
74,550	**74,600**	18,386	15,811	19,308	16,819
74,600	**74,650**	18,402	15,825	19,326	16,833
74,650	**74,700**	18,417	15,839	19,344	16,847
74,700	**74,750**	18,433	15,853	19,362	16,861
74,750	**74,800**	18,448	15,867	19,380	16,875
74,800	**74,850**	18,464	15,881	19,398	16,889
74,850	**74,900**	18,479	15,895	19,416	16,903
74,900	**74,950**	18,495	15,909	19,434	16,917
74,950	**75,000**	18,510	15,923	19,452	16,931
75,000					
75,000	**75,050**	18,526	15,937	19,470	16,945
75,050	**75,100**	18,541	15,951	19,488	16,959
75,100	**75,150**	18,557	15,965	19,506	16,973
75,150	**75,200**	18,572	15,979	19,524	16,987
75,200	**75,250**	18,588	15,993	19,542	17,001
75,250	**75,300**	18,603	16,007	19,560	17,015
75,300	**75,350**	18,619	16,021	19,578	17,029
75,350	**75,400**	18,634	16,035	19,596	17,043
75,400	**75,450**	18,650	16,049	19,614	17,057
75,450	**75,500**	18,665	16,063	19,632	17,071
75,500	**75,550**	18,681	16,077	19,650	17,085
75,550	**75,600**	18,696	16,091	19,668	17,099
75,600	**75,650**	18,712	16,105	19,686	17,113
75,650	**75,700**	18,727	16,119	19,704	17,127
75,700	**75,750**	18,743	16,133	19,722	17,141
75,750	**75,800**	18,758	16,147	19,740	17,155
75,800	**75,850**	18,774	16,161	19,758	17,169
75,850	**75,900**	18,789	16,175	19,776	17,183
75,900	**75,950**	18,805	16,189	19,794	17,197
75,950	**76,000**	18,820	16,203	19,812	17,211
76,000					
76,000	**76,050**	18,836	16,217	19,830	17,225
76,050	**76,100**	18,851	16,231	19,848	17,239
76,100	**76,150**	18,867	16,245	19,866	17,253
76,150	**76,200**	18,882	16,259	19,884	17,267
76,200	**76,250**	18,898	16,273	19,902	17,281
76,250	**76,300**	18,913	16,287	19,920	17,295
76,300	**76,350**	18,929	16,301	19,938	17,309
76,350	**76,400**	18,944	16,315	19,956	17,323
76,400	**76,450**	18,960	16,329	19,974	17,337
76,450	**76,500**	18,975	16,343	19,992	17,351
76,500	**76,550**	18,991	16,357	20,010	17,365
76,550	**76,600**	19,006	16,371	20,028	17,379
76,600	**76,650**	19,022	16,385	20,046	17,393
76,650	**76,700**	19,037	16,399	20,064	17,407
76,700	**76,750**	19,053	16,413	20,082	17,421
76,750	**76,800**	19,068	16,427	20,100	17,435
76,800	**76,850**	19,084	16,441	20,118	17,449
76,850	**76,900**	19,099	16,455	20,136	17,463
76,900	**76,950**	19,115	16,469	20,154	17,477
76,950	**77,000**	19,130	16,483	20,172	17,491

* This column must also be used by a qualifying widow(er).

Continued on next page

EXHIBIT 18-3 (continued)

1995 Tax Rate Schedules

Caution: *Use* ***only*** *if your taxable income (Form 1040, line 37) is* ***$100,000*** *or more. If less, use the* ***Tax Table.*** *Even though you cannot use the tax rate schedules below if your taxable income is less than $100,000, all levels of taxable income are shown so taxpayers can see the tax rate that applies to each level.*

Schedule X—Use if your filing status is **Single**

If the amount on Form 1040, line 37, is: *Over—*	*But not over—*	Enter on Form 1040, line 38	*of the amount over—*
$0	$23,350	.... 15%	**$0**
23,350	56,550	**$3,502.50 + 28%**	**23,350**
56,550	117,950	**12,798.50 + 31%**	**56,550**
117,950	256,500	**31,832.50 + 36%**	**117,950**
256,500	....	**81,710.50 + 39.6%**	**256,500**

Schedule Y-1—Use if your filing status is **Married filing jointly** or **Qualifying widow(er)**

If the amount on Form 1040, line 37, is: *Over—*	*But not over—*	Enter on Form 1040, line 38	*of the amount over—*
$0	$39,000	.... 15%	**$0**
39,000	94,250	**$5,850.00 + 28%**	**39,000**
94,250	143,600	**21,320.00 + 31%**	**94,250**
143,600	256,500	**36,618.50 + 36%**	**143,600**
256,500	....	**77,262.50 + 39.6%**	**256,500**

Schedule Y-2—Use if your filing status is **Married filing separately**

If the amount on Form 1040, line 37, is: *Over—*	*But not over—*	Enter on Form 1040, line 38	*of the amount over—*
$0	$19,500	.... 15%	**$0**
19,500	47,125	**$2,925.00 + 28%**	**19,500**
47,125	71,800	**10,660.00 + 31%**	**47,125**
71,800	128,250	**18,309.25 + 36%**	**71,800**
128,250	....	**38,631.25 + 39.6%**	**128,250**

Schedule Z—Use if your filing status is **Head of household**

If the amount on Form 1040, line 37, is: *Over—*	*But not over—*	Enter on Form 1040, line 38	*of the amount over—*
$0	$31,250	.... 15%	**$0**
31,250	80,750	**$4,687.50 + 28%**	**31,250**
80,750	130,800	**18,547.50 + 31%**	**80,750**
130,800	256,500	**34,063.00 + 36%**	**130,800**
256,500	....	**79,315.00 + 39.6%**	**256,500**

EXHIBIT 18-4 TAX RATE SCHEDULES

Tax Rate Schedule

If taxable income (line 30, Form 1120, or line 26, Form 1120-A) on page 1 is:

Over—	*But not over—*	*Tax is:*	*Of the amount over—*
$0	$50,000	**15%**	$0
50,000	75,000	**$7,500 + 25%**	50,000
75,000	100,000	**13,750 + 34%**	75,000
100,000	335,000	**22,250 + 39%**	100,000
335,000	10,000,000	**113,900 + 34%**	335,000
10,000,000	15,000,000	**3,400,000 + 35%**	10,000,000
15,000,000	18,333,333	**5,150,000 + 38%**	15,000,000
18,333,333	- - - - -	**35%**	0

EXHIBIT 18-5

CORPORATE TAX RATE SCHEDULE

	Term Insurance				Permanent Insurance					
	5-Year Term		10-Year Term		Whole Life		20-Payment Life		20-Year Endowment	
Age	**Male**	**Female**	**Male**	**Female**	**Male**	**Female**	**Male**	**Female**	**Male**	**Female**
18	$2.32	$1.90	$4.33	$4.01	$13.22	$11.17	$23.14	$19.21	$33.22	$29.12
19	2.38	1.96	4.42	4.12	13.60	11.68	24.42	20.92	33.68	30.04
20	2.43	2.07	4.49	4.20	14.12	12.09	25.10	21.50	34.42	31.28
21	2.49	2.15	4.57	4.29	14.53	12.53	25.83	22.11	34.90	31.79
22	2.55	2.22	4.64	4.36	14.97	12.96	26.42	22.89	35.27	32.40
23	2.62	2.30	4.70	4.42	15.39	13.41	27.01	23.47	35.70	32.93
24	2.69	2.37	4.79	4.47	15.90	13.92	27.74	24.26	36.49	33.61
25	2.77	2.45	4.85	4.51	16.38	14.38	28.40	25.04	37.02	34.87
26	2.84	2.51	4.92	4.60	16.91	14.77	29.11	25.96	37.67	35.30
27	2.90	2.58	5.11	4.69	17.27	15.23	29.97	26.83	38.23	35.96
28	2.98	2.64	5.18	4.77	17.76	15.66	30.68	27.54	38.96	36.44
29	3.07	2.70	5.23	4.84	18.12	16.18	31.52	28.09	39.42	37.21
30	3.14	2.78	5.30	4.93	18.54	16.71	32.15	28.73	40.19	37.80
35	3.43	2.92	6.42	5.35	24.19	22.52	37.10	33.12	43.67	39.19
40	4.23	3.90	7.14	6.24	27.21	25.40	42.27	36.29	48.20	42.25
45	6.12	5.18	8.81	7.40	33.02	29.16	48.73	39.08	51.11	46.04
50	9.72	8.73	14.19	9.11	37.94	33.57	56.31	44.16	58.49	49.20
55	16.25	12.82	22.03	13.17	45.83	37.02	61.09	49.40	71.28	53.16
60	24.10	19.43	37.70	24.82	53.98	42.24	70.43	52.55	79.15	58.08

TABLE 19-1

ANNUAL LIFE INSURANCE PREMIUMS (PER $1,000 OF FACE VALUE)

Premium Paid	**Percent of Annual Premium**
Semiannually	52%
Quarterly	26%
Monthly	9%

TABLE 19-2

LIFE INSURANCE—PREMIUM FACTORS

End of Year	Whole Life Option 1 Cash Value	Whole Life Option 2 Reduced Paid-up Insurance	Whole Life Option 3 Extended Term Years	Whole Life Option 3 Extended Term Days	20-Payment Life Option 1 Cash Value	20-Payment Life Option 2 Reduced Paid-up Insurance	20-Payment Life Option 3 Extended Term Years	20-Payment Life Option 3 Extended Term Days	20-Year Endowment Option 1 Cash Value	20-Year Endowment Option 2 Reduced Paid-up Insurance	20-Year Endowment Option 3 Extended Term Years	20-Year Endowment Option 3 Extended Term Days
3	$11	$25	2	17	$29	$90	4	217	$39	$97	7	132
5	32	64	9	23	73	212	14	86	91	233	19	204
7	54	99	13	142	101	367	23	152	186	381	26	310
10	98	186	17	54	191	496	30	206	324	512	32	117
15	157	314	21	218	322	789	34	142	647	794	37	350
20	262	491	25	77	505	1,000	-Life-		1,000	1,000	-Life-	

TABLE 19-3

NONFORFEITURE OPTIONS (PER $1,000 OF FACE VALUE ISSUED TO FEMALE AT AGE 20)

Area Rating	Structural Classification A Building	A Contents	B Building	B Contents	C Building	C Contents	D Building	D Contents
1	$.21	$.24	$.32	$ .37	$ .38	$ .42	$ .44	$ .48
2	.38	.42	.39	.48	.43	.51	.57	.69
3	.44	.51	.55	.66	.69	.77	.76	.85
4	.59	.68	.76	.83	.87	1.04	.98	1.27
5	.64	.73	.92	1.09	1.08	1.13	1.39	1.43

TABLE 19-4

ANNUAL FIRE INSURANCE PREMIUMS (PER $100 OF FACE VALUE)

Time Policy Is in Force	Percent of Annual Premium	Time Policy Is in Force	Percent of Annual Premium
5 days	8%	4 months	50%
10 days	10	5 months	60
15 days	14	6 months	70
20 days	16	7 months	75
25 days	18	8 months	80
		9 months	85
1 month	20	10 months	90
2 months	30	11 months	95
3 months	40	12 months	100

TABLE 19-5

PROPERTY INSURANCE SHORT-RATE SCHEDULE

Territory	Driver Class	Bodily Injury (000)					Property Damage (000)				
		10/20	15/30	25/50	50/100	100/300	5	10	25	50	100
1	1	$61	$73	$88	$92	$113	$46	$49	$53	$58	$64
	2	63	75	81	94	116	48	51	55	61	66
	3	65	78	84	98	118	52	54	58	63	69
	4	69	81	86	101	121	54	56	60	65	71
2	1	66	75	83	93	114	56	63	68	73	77
	2	69	77	88	98	117	58	64	70	75	79
	3	75	82	92	104	119	59	66	71	76	82
	4	78	86	95	109	122	62	67	73	78	84
3	1	73	77	84	95	116	64	65	72	76	81
	2	78	83	86	99	119	66	69	74	80	83
	3	84	88	92	103	124	70	73	77	82	85
	4	87	93	95	106	128	72	78	81	85	89
4	1	77	81	86	99	118	76	78	83	88	92
	2	81	86	93	103	121	79	83	87	91	95
	3	87	92	100	106	126	80	84	88	93	97
	4	90	94	103	111	132	84	86	91	94	100

TABLE 19-6

MOTOR VEHICLE LIABILITY INSURANCE PREMIUMS
ANNUAL—BODILY INJURY AND PROPERTY DAMAGE RATES

Model Class	Vehicle Age	Territory 1 & 2				Territory 3 & 4			
		Collision		Comprehensive		Collision		Comprehensive	
		$250 Deductible	$500 Deductible	Full Coverage	$100 Deductible	$250 Deductible	$500 Deductible	Full Coverage	$100 Deductible
A–G	0–1	$89	$81	$63	$59	$95	$88	$67	$61
	2–3	87	79	60	57	93	84	63	58
	4–5	86	77	58	54	89	81	60	57
	6+	84	76	55	50	86	78	57	52
H–L	0–1	96	92	78	71	104	95	83	75
	2–3	93	89	76	68	101	90	80	72
	4–5	89	85	74	66	96	87	78	68
	6+	86	81	70	64	92	84	74	66
M–R	0–1	108	104	86	83	112	106	91	88
	2–3	104	101	83	79	109	104	88	82
	4–5	100	98	79	75	104	101	84	77
	6+	94	90	75	71	100	96	80	74
S–Z	0–1	120	115	111	108	124	116	119	113
	2–3	116	112	106	104	121	114	115	109
	4–5	111	107	101	99	116	110	111	106
	6+	108	103	98	96	111	107	108	101

TABLE 19-7

MOTOR VEHICLE INSURANCE PREMIUMS
ANNUAL—COLLISION AND COMPREHENSIVE RATES

EXHIBIT 20-3

THE WALL STREET JOURNAL
NEW YORK STOCK EXCHANGE
COMPOSITE TRANSACTIONS

	52 Weeks Hi	Lo	Stock	Sym	Div	Yld %	PE	Vol 100s	Hi	Lo	Close	Net Chg
▼	48½	41¼	JamesRiver pfL		3.50	8.6	...	84	41½	40½	40½	– ¾
	26¾	22¾	JamesRiver pfO		2.06	9.0	...	59	23⅛	22¾	23	...
	2⅞	7/32	vjJamesway	JMY		...	...	147	⅝	½	9/16	...
	18	12	JapanEquity	JEQ	1.09e	6.6	...	351	16⅝	16⅜	16½	...
	14½	8¼	JapanOTC	JOF		...	...	504	13	12¾	12¾	– ¼
	26¾	16⅝	JardineFlemg	JFC	1.73e	9.8	...	327	17⅞	17¼	17⅝	– ⅜
n	15¼	13⅞	JardineIndia	JFI		...	...	170	14⅜	14	14⅛	+ ⅛
	57⅞	43⅜	JeffPilot	JP	1.72f	3.5	13	453	49⅞	49¼	49¾	+ ½
	17⅜	5⅝	JennyCraig	JC	.60	9.8	6	129	6¼	6⅛	6⅛	– ⅛
▼	61½	51½	JerCentl pf		4.00	7.8	...	z20	51	51	51	–1
	42⅛	18⅜	JohnAldenFnl	JA	.40f	1.1	11	902	35⅞	34⅞	35⅜	+ ⅜
	41⅝	19⅜	JohnNuveen	JNC	.64f	3.1	12	2131	20¾	20	20¾	+ ½
	45¾	35⅝	JohnsJohns	JNJ	1.04	2.6	15	14264	40⅜	38⅝	40⅜	+1½
	61¾	45½	JohnsContrl	JCI	1.44	2.8	15	831	52⅜	50⅝	51¾	+1¼
s	13¾	8¼	JohnstnInd	JII	.33	3.3	13	42	10	10	10	– ¼
	35¾	18⅝	JonesApparel	JNY		...	17	949	32⅛	32	32	– ¼
	27⅛	16¼	Jostens	JOS	.88	5.3	dd	425	16¾	16½	16⅝	+ ⅛
	14	8½	JoyTech	JOY		...	29	471	12¾	12⅝	12¾	+ ¼
	15	12¼	JundtGrowFd	JF	.53e	3.9	...	323	13¾	13½	13¾	+ ¼
			-K-K-K-									
s	32¾	17⅛	KCS Engy	KCS	.08	.3	29	216	25⅞	24⅝	24⅞	– ⅞
	27¼	13½	KLM	KLM		...	...	628	26½	26⅛	26⅜	+ ⅛
s▼	30	21⅜	KN Engy	KNE	.96	4.5	14	60	21¾	21¼	21¼	– ½
	32¾	25⅞	KU Engy	KU	1.64f	6.2	12	212	26⅝	26⅛	26⅝	+ ¼
	28⅞	26⅝	KIII pf		2.88	10.6	...	5	27¼	27¼	27¼	+ ¼
	12½	6⅜	KaiserAlum	KLU		...	dd	440	8⅞	8¾	8¾	– ⅛
n▼	8¾	7¼	KaiserAlum pfA		.65	8.8	...	128	7⅜	7	7⅜	...
n▼	12⅛	9⅝	KaiserAlum pfD		.12p	1.2	...	144	9¾	9½	9⅝	...
	28⅜	22⅞	KanebPipe	KPP	2.20	8.7	...	86	25¾	25⅜	25⅜	...
	4⅛	2⅝	KanebSvcs	KAB		...	7	410	3⅜	3¼	3¼	– ⅛
	26¼	20⅝	KanCityPL	KLT	1.48	6.8	14	794	21¾	21¼	21¾	+ ⅜
	52⅝	**30½**	**KanCitySou**	**KSU**	**.30**	**.7**	**19**	**2503**	**44⅜**	**41⅜**	**41¾**	**–2¾**
	16⅝	14	KanCitySou pf		1.00	6.9	4	z10	14½	14½	14½	– ½
	10¾	7	KaslerHldg	KAS	.05p	.7	26	372	7⅜	7¼	7¼	...
	28⅛	25	KatyInd	KT	.25	1.0	dd	9	25⅜	25¼	25⅜	+ ⅜

	1	2	3	4	5	6	7	8	9	10	11	12
	52 Weeks Hi	**Lo**	**Stock**	**Sym**	**Div**	**Yld %**	**PE**	**Vol 100s**	**Hi**	**Lo**	**Close**	**Net Chg**
	1⅜	½	MLawrencEds	MLE		...	dd	10	1	15/16	1	...
s	47¼	35¾	MartinMar	ML	.90	2.0	10	1857	44½	44	44¼	+ ¼
n▼	25⅞	20½	MartinMarMatl	MLM		...	...	231	20½	20	20⅛	– ⅜
s	35¾	12⅜	MarvelEntn	MRV		...	29	2207	16⅛	15½	16⅛	+ ¼
	39⅞	25½	Masco	MAS	.68	2.4	19	7359	28¾	27¾	28	...
	28⅛	**16⅛**	**Mascotech**	**MSX**	**.08**	**.4**	**31**	**1821**	**20⅝**	**19½**	**19½**	**–1⅛**
n	25⅜	19	Mascotech pf		1.20	6.2	...	29	19⅝	19¼	19¼	– ⅝
	33¼	28⅛	MassMuInv	MCI	2.80	8.9	...	36	31⅝	31⅜	31⅜	+ ¼
	8¾	7½	MassMuPrtInv	MPV	.68	8.9	...	131	7⅞	7⅝	7⅝	– ¼
	26⅝	17	MaterlSci	MSC		...	22	89	22⅛	21¾	22⅛	+ ¼
	16½	9¼	MatlackSys	MLK		...	20	51	15⅝	15	15⅝	+ ½
	176	114	MatsuElec	MC	1.27e	.8	...	8	167	167	167	...
s	26⅞	18¼	Mattel	MAT	.24	1.0	26	7958	24¾	23⅝	23⅞	– ¼
	5¼	**3⅞**	**MaunaLoa**	**NUT**	**.40**	**9.1**	**31**	**91**	**4⅝**	**4⅜**	**4⅜**	**– ¼**
	10⅜	4⅛	MaxusEngy	MXS		...	dd	2819	4⅝	4⅜	4½	...
n	25¾	21⅛	MaxusEngy pfA		2.50	11.4	...	17	21⅞	21¾	21⅞	+ ⅛
	49⅞	39	MaxusEngy pf		4.00	10.2	...	5	39⅜	39¼	39¼	– ⅛
	20⅝	12¼	MaxximMed	MAM		...	16	16	16¼	16⅛	16¼	...
s	46½	33⅜	MayDeptStrs	MA	1.04f	2.6	14	5448	41¼	39½	39⅝	– ¾
	31½	16¾	Maybelline	MAY	.24	.8	24	367	31¼	30⅝	31⅛	+ ⅜
	20	13	Maytag	MYG	.50	2.9	36	3013	17⅝	17⅛	17⅛	...
n	28¼	22⅝	McArthurGln	MCG	.64e	2.7	...	256	24	23⅞	23⅞	– ⅛
	25⅝	18⅛	McClatchy A	MNI	.32f	1.5	20	42	22	21⅝	22	+ ⅛
	32⅞	19⅞	McDermInt	MDR	1.00	4.7	12	1786	21¼	20¾	21⅛	+ ¼
	35⅞	29	McDermInt pfA		2.20	7.5	...	38	29⅝	29¼	29½	+ ⅛
	31⅞	30⅞	McDermInt pfB		2.60	8.3	...	8	31¼	31¼	31¼	+ ⅛
s	16⅞	12¼	McDnInvst	MDD	.30	2.1	6	69	14⅛	13⅞	14	– ¼
	62½	45½	McDonalds	MCD	.43	.8	19	5373	57¼	56⅜	56⅞	+ ⅜
	28	24½	McDonalds dep pf		1.93	7.7	...	3138	24 61/64	24⅝	24 61/64	...
	122⅝	58	McDonDoug	MD	1.40	1.2	14	2221	115⅛	113	113⅞	– ¼
	75¼	55¼	McGrawH	MHP	2.32f	3.5	cc	785	65⅞	64¾	65⅜	+ ⅜
	68½	38⅝	McKesson	MCK	1.68	2.6	18	997	64¼	62⅞	63⅞	+1⅛
n	17¾	13¼	McWhortrTch wi			...	...	21	17⅜	17¼	17⅜	...
	48½	39⅛	Mead	MEA	1.00	2.5	19	974	40⅜	39½	39⅝	– ¾
	21	15¾	Measurx	MX	.44	2.4	38	45	18¾	18⅛	18⅛	– ¾
s	10¼	5⅝	MedChmPrd	MCH	1.05t	17.9	11	135	6	5⅞	5⅞	...

CORPORATION BONDS
Volume, $30,630,000

Bonds	Cur Yld	Vol	Close	Net Chg.
AMR 9s16	9.1	95	99	− ⅜
AMR 8.10s98	7.9	7	102	− ⅝
ATT 4¾98	5.1	10	94	+ ¾
ATT 4⅜96	4.5	55	96¾	+ ½
ATT 4⅜99	4.8	42	90½	+ ¼
ATT 6s00	6.3	79	96	+ ⅜
ATT 5⅛01	5.6	21	91½	+ ½
ATT 8⅝31	8.2	10	105¾	+ 1¼
ATT 7⅛02	7.1	543	100⅜	...
ATT 8⅛22	8.0	289	101⅝	+ ⅛
ATT 8⅛24	8.0	65	101⅞	+ ½
ATT 4½96	4.6	103	98¼	+ ⅜
ATT 6¾04	7.0	691	96	− ⅛
Actava 9⅞97	10.1	15	97¾	− ⅛
Actava 9½98	10.3	114	92½	+ 1¼
Advst 9s08	cv	9	100¾	+ 2¾
AirbF 6¾01	cv	5	104	...
AlskAr 6⅞14	cv	8	81	+ ⅝
AlskAr zr06	...	108	38¾	...
AlbnyInt 5s02	cv	19	88½	+ ½
AlegCp 6½14	cv	10	98	− 1
Allwst 7¼14	cv	15	87⅞	− ⅛
AForP 5s30	7.9	1	63½	+ ⅛
AAir dc6¼96	6.3	15	99½	...
AmBrnd 8½03	8.3	5	102⅞	− ⅝
AmBrnd 8⅝21	8.4	15	103	− 7
Ametek 9¾04	9.8	34	100	+ 1
Anhr 8⅝16	8.5	14	101¾	− 1¼
AnnTaylr 8¾00	9.2	75	95¼	− ⅛
AshO 6¾14	cv	17	99¾	+ ¼
ARch 10⅜95	9.9	28	105	− ½
AutDt zr12	...	20	39⅜	+ ⅞
Avnet 6s12	cv	3	101¾	+ 2¼
Barnt 8½99	8.1	25	105⅛	+ ¼
BarBks 10⅞03	9.5	15	115	...
BellPa 7⅛12	7.6	4	93½	+ ¼
BellsoT 8¼32	8.1	10	102⅛	+ ¼
BellsoT 7⅞32	8.0	370	98	− ¼
BellsoT 6½00	6.6	70	99	...
BellsoT 7½33	8.0	100	94	+ ½
Bellso 5⅞09	6.7	84	88	+ 2
BstBuy 8⅝00	9.0	244	95¾	− ⅛
BethSt 9s00	9.0	19	100⅜	− ⅛
BethSt 8.45s05	8.9	29	94¾	− ¼
Bevrly 7⅝03	cv	26	100½	− ¾
BlkBst 6⅝98	6.8	7	97½	+ 1⅝
Boeing 8⅜96	7.9	10	105⅜	+ 1⅛
BoisC 7s16	cv	57	87½	− ½
Bordn 8⅜16	8.6	25	97½	+ 1½
BorgWS 9⅛03	9.5	10	96	+ ½
BwnSh 9¼05	cv	11	99¼	...

1	2	3	4	5
Bonds	Cur Yld	Vol	Close	Net Chg.
CnPw 6⅞98	7.0	5	97¾	− ¾
viCtlInf 9s06f	cv	45	9	+ ½
Coopr 10⅝05f	...	11	70	− ⅛
CrayRs 6⅛11	cv	10	83½	...
DataGn 01	cv	39	84½	− ¼
Datpnt 8⅞06	cv	27	76	+ 1
DaytP 8s03	8.0	7	100⅜	− ⅝
DeereCa 7.2s97	7.1	10	100⅞	− 2⅛
DetEd 6.4s98	6.5	8	98¾	− 1¾
DiaSTel 7s08	7.1	5	98½	− ½
Dole 7s03	7.6	15	92½	− ¾
duPnt dc6s01	6.5	352	92¾	− ¼
EMC 4¼01	cv	15	105	...
Eckerd 9¼04	9.9	749	93	+ ½
EmbSuit 11s99	10.6	9	104	...
EngStr zr97	...	3	74⅞	− ⅛
Enron 10¾98	10.2	2	105¾	+ ¾
EthAln 8¾01	9.0	30	97½	...
Exxon 8¼94	8.1	10	102¼	+ 1 5/32
F&M 11½03	12.3	11	93⅛	− 1¾
FairCp 12s01	12.1	2	99	+ 1
FairCp 13⅛06	13.4	12	98	...
Fldcst 6s12	cv	56	89	− ¼
FreptM 10⅞01	10.1	10	107½	− 2
Frpt dc6.55s01	cv	35	89	+ 1¼
FMRP 8¾04	9.1	5	96	+ 2
GMA dc6s11	7.5	100	79¾	− ¾
GMA zr12	...	60	218⅞	− ⅛
GMA zr15	...	50	178¼	+ 1½
GMA 8¼16	8.4	135	98⅝	+ ⅝
GMA 8s94	7.9	5	101 5/16	− 3/16
GMA 8⅜97	8.0	10	104⅞	+ 1⅛
GPA Del 8¾98	10.6	173	82⅝	...
GnCorp 8s02	7.7	2	104	...
GEICap 7⅞06	7.6	10	103½	+ ⅛
GHost 11½02	11.7	210	98½	− ⅜
GHost 8s02	cv	73	86	+ 1
GnSgnl 5¾02	cv	31	100	− 1
Gene 10⅜03	12.1	85	86	+ 1
GenesisH 6s03	cv	25	114	+ 1¼
Genrad 7¼11	cv	1	78	+ 1
GaGlf 15s00	14.1	700	106¾	− ½
GaPw 10s16J	9.8	95	102½	− ½
Getty 14s00	...	13	102 25/32	...
Grancre 6½03	6.6	10	98	− ½
Hallb zr06	...	1	48½	− ¼
Hallwd na13½09	...	50	88½	...
Hallw na13½09B	...	2	90	+ ⅜
Hallwd 7s00	10.6	36	65⅞	+ 1⅞
Hlttrst 8¾05	9.6	561	91½	− ¼
HomeDp 4½97	cv	14	113¼	− 1¼
HostM 10⅝00	10.5	21	101	...
HostM 9⅛00	9.5	147	96¼	− ½
HostM 9⅞01	10.1	75	97⅝	− ⅜

EXHIBIT 20-5

THE WALL STREET JOURNAL
NEW YORK EXCHANGE BONDS

Practice Problems

CHAPTER 1 PRACTICE EXERCISES

1. Sandy Sullivan started the day with $84 in her purse. She spent $21 at the dry cleaners, $16 at the grocery store, and $18 for gasoline. On the way home, she cashed a check for $40.

   a. How much money does she have at the end of the day?

   b. If Sandy has a spending budget limit of $50, is she over or under her budget?

   c. By how much is she over or under her budget?

2. The following chart shows the meals served per day by the Riverside Restaurant. Total each column to get the daily totals. Total each row to get the total meals per category. What is the week's grand total?

	Monday	Tuesday	Wednesday	Thursday	Friday	Total Meals
Breakfast	52	65	32	57	72	______
Lunch	98	68	41	39	97	______
Dinner	56	95	72	108	112	______
Late Night	208	148	169	214	303	______
Daily Totals	___	___	___	___	___	**Grand Total** ______

3. A company with 245 employees has $364,070 in profit sharing to distribute. As the finance manager, calculate how much each employee should receive.

4. At the beginning of the fall semester, a college sold 12,488 lower-deck football tickets at $15 each; 6,432 upper-level tickets at $12 each; and 6,900 end zone tickets at $9 each.

   a. How many tickets were sold?

b. What was the total dollar amount of the ticket sales?

5. The Arrow Hardware had sales of $18,630 last week. The store had 690 transactions.

   a. What was the average dollars per transaction?

   b. If the store was open 6 days, what was the average dollar sales per day?

6. Expenses of a company for March are $3,590 for salaries, $767 for electricity, $235 for insurance, $2,388 for advertising, $1,200 for rent, and $1,875 for miscellaneous expenses.

   a. What are the total expenses for March?

   b. If the March total is representative of a typical month, what are the total expenses for this company per year?

7. Albrecht Realty, Inc., sells four parcels of land for $32,500 each; one farm for $55,600; and two homes, one for $79,200, and one for $96,700. What are the company's total sales?

8. Jim Thompson bought 500 shares of stock at $51 per share. Five months later he sold the 500 shares at $77 per share. If the total stockbroker's commission was $800, how much profit did Thompson make?

9. A company projected profits to be $14,560,000. If actual profits were only $12,339,590, by how much was the projection over the actual profits?

1. Sue Anderson started a diet weighing $134\frac{3}{4}$. The first week she lost $2\frac{1}{4}$ pounds, the second week she lost $1\frac{1}{2}$ pounds, and the third week she gained $1\frac{1}{8}$ pounds. How much does Sue weigh now?

2. Stan Edwards increased his earnings by $\frac{1}{5}$ from last year. If he made $25,500 last year, how much *more* did he make this year?

3. Pam Singer earns $140 per day. On Tuesday she only worked $\frac{4}{7}$ of a day. How much did she earn?

4. Fancy Fruit Wholesalers purchases 350 crates of apples from the Sun-Ripe Orchard. They intend to repack the apples in smaller boxes to be shipped to supermarkets. If each box contains $\frac{5}{8}$ of a crate, how many boxes can be packed?

5. Will Peterson, a farmer, sold $2\frac{3}{5}$ acres of his $18\frac{1}{4}$ acres. How many acres does he have left?

6. Fleming's Warehouse contains 19,667 square feet. If each storage bin requires $35\frac{1}{2}$ square feet, how many bins can be stored in the warehouse?

7. Alice McEnroe has a recipe for noodle pudding that serves four people. Alice wants to make the pudding for a dinner party, serving ten people. Calculate the new amount for each ingredient. Hint: 10 people is $2\frac{1}{2}$ times as many as 4 people. ($\frac{10}{4} = 2\frac{1}{2}$)

Ingredient	Quantity (4 people)	Quantity (10 people)
cooked noodles	24 ounces	________
sour cream	$2\frac{4}{5}$ cups	________
sugar	$\frac{2}{3}$ cup	________
pitted cherries	$7\frac{1}{2}$ ounces	________
vanilla	$\frac{3}{4}$ teaspoon	________
eggs	2 eggs	________

8. Three-fifths of the cars in the school parking lot today were made by General Motors. If 4,800 cars are in the lot, how many are GM?

9. Sally Wilder worked $6\frac{1}{5}$ hours on Monday, $8\frac{1}{2}$ hours on Tuesday, and $11\frac{1}{4}$ hours on Wednesday. If she intends to work 35 hours this week, how many more hours must Sally work?

10. A recent market research survey at Winn Dixie showed that $\frac{5}{8}$ of the people interviewed preferred double coupons rather than trading stamps as a supermarket promotion. If 1,800 shoppers were interviewed in Toledo, Ohio, how many preferred trading stamps? Hint: $\frac{8}{8}$ represents all the shoppers. Since $\frac{5}{8}$ preferred coupons, $\frac{3}{8}$ preferred stamps ($\frac{8}{8} - \frac{5}{8} = \frac{3}{8}$).

11. What is the total price of 1,600 shares of stock, selling for $\$3\frac{5}{8}$ per share?

12. Don Bailey bought stock at $\$54\frac{1}{4}$. The first week the stock went up $\$7\frac{7}{8}$, but the following week it fell $\$9\frac{3}{4}$. What is the new price of the stock?

13. A union takes $\frac{1}{25}$ of John's wages as dues. If he made \$825 last week, how much was deducted for dues?

14. Eddie Quick can assemble $6\frac{4}{5}$ chairs per hour. Warren Weary can produce $4\frac{1}{2}$ per hour. In a ten-hour shift, how many more chairs does Eddie make than Warren?

15. How many pieces of rope, $18\frac{3}{4}$ feet long, can be cut from a roll measuring 1,500 feet?

16. Sears advertises $\frac{1}{3}$ off Craftsman 12-drawer tool chests. If the original price is \$360, what is the sale price of the chest?

17. A lottery of \$260,000 is split among 65 winners. Each will receive $\frac{1}{65}$ of the prize. How much is each share of the winnings?

18. A boat hull weighs $1{,}658\frac{1}{2}$ pounds, and the outboard motor weighs $645\frac{1}{4}$ pounds. A standard trailer has a weight limit of 2,100 pounds.

a. Is this a safe combination, or will a heavy-duty trailer be necessary?

b. By how many pounds is the load over or under the limit?

19. Production line workers at the Holsum Bakery use $10\frac{1}{2}$ ounces of flour for each loaf of bread they bake. If a worker used 6,300 ounces of flour last week, how many loaves were baked?

1. A Ferrari began an auto race with a full tank of fuel totaling 21.5 gallons. During the two pit stops, the crew added 17.85 gallons and 14.32 gallons. After finishing, the car had 5.1 gallons left in the tank. How many gallons were used during the race?

2. A riding lawnmower is on sale this week at Sears for $1,854.97.

   a. If the original price was $2,435.50, how much can be saved by purchasing the lawnmower on sale?

   b. If you purchase an extended service policy for $224.16 and the tax is $127.84, what is the total purchase price of the lawnmower on sale?

3. Craig Adams makes $9.57 per hour. If he worked 38 hours last week, how much did he earn?

4. Pam Fullerton bought 3.6 pounds of sirloin steak at $5.99 per pound; 6.8 pounds of potatoes at $.89 per pound; and .61 pound of spinach at $.37 per pound. What was the total cost of the food?

5. Willie Jordan earned $537.08 last week. If he worked 46.3 hours, what is his hourly wage?

6. Bryant Mathis ran a 10K race in 49.6 minutes. The term 10K means 10 kilometers and equals 6.2 miles.

   a. How many minutes per mile did Bryant run?

   b. If the winner finished in 37.1 minutes, how much longer did it take Bryant to finish?

7. Fabulous Furniture had a year-end clearance sale last week. At the sale, Jenny Jones purchased a sofa for $588.50, marked down from $767.99; and 2 chairs for $99.68 each, marked down from $159.95 each. In addition, she purchased a wall unit for $639.77, which is one-half the original price.

   a. What was the total of Jenny's purchases?

   b. How much did she save?

8. You are the payroll manager for the Rand Company. Rounding to dollars and cents, complete the following payroll data sheet:

Employee	Hours	Hourly Wage	Total	Deductions	Net Pay
Williams	37.6	$6.80	______	$51.68	______
Johnson	43.22	$8.25	______	$102.64	______
Calder	39.37	$10.47	______	______	$287.95
Sanford	27.7	______	$489.61	$122.63	______

9. A shopping center received $416,500 rent from its tenants last month. The center charges $2.45 per square foot.

   a. How many square feet is the shopping center?

b. If maintenance costs are $.028 per square foot per month, what is the total cost of maintenance per month?

10. A carpenter spends $18.85 on nails. If nails sell for $.65 per pound, how many pounds did the carpenter purchase?

11. The Academy Lumber Company has delivery orders of 145 sheets of plywood, each weighing 18.6 pounds; and 64 bags of concrete mix, each weighing 69.7 pounds. Academy's small delivery truck has a load capacity of 7,000 pounds.

a. Can this order be shipped in one truckload?

b. By how much is the order over or under the weight limit?

12. Ben Elderd buys a farm for $146,575. If the farm has $112\frac{3}{4}$ acres, what is the cost per acre?

13. A ship has a cargo area of 10,000 cubic feet. How many $18\frac{2}{5}$-cubic-foot storage containers can the ship hold? Round to the nearest whole container.

14. **a.** What is the cost of 2,000 shares of stock selling at $\$9\frac{5}{8}$ per share?

b. If the broker's commission is $458.23, what is the total cost of the transaction?

1. You are the bookkeeper for Ullman & Company. In proper form, fill out the checks, check stubs, and deposit slip, based on the following information:
   a. The balance brought forward is $4,599.36.
   b. On May 26, 19xx, a check in the amount of $1,875.40 is written to The Walton Group, for supplies.
   c. On June 4, a deposit is made of checks totaling $5,690.11, currency of $450.00, and coins of $66.75.
   d. On June 9, a check in the amount of $1,391.88 is written to Central Supply, Inc., for roofing materials.

IF TAX DEDUCTIBLE CHECK HERE ☐ $______

3088

______ 19 ____

TO ______

FOR ______

	DOLLARS	CENTS
BAL. FWD.		
DEPOSIT		
DEPOSIT		
TOTAL		
THIS ITEM		
SUB-TOTAL		
OTHER DEDUCT. (IF ANY)		
BAL. FWD.		

Ullman & Company
390 Cherry Rd.
Boise, Idaho 54948

3088

63-398/670

______ 19 ____

PAY TO THE ORDER OF ______ $ ______

______ DOLLARS

Barnett Bank
037-049
11755 Biscayne Blvd.
North Miami, Florida 33161

FOR ______

⑆067003985⑆ 3088 630211502 ⑈

IF TAX DEDUCTIBLE CHECK HERE ☐ $______

3089

______ 19 ____

TO ______

FOR ______

	DOLLARS	CENTS
BAL. FWD.		
DEPOSIT		
DEPOSIT		
TOTAL		
THIS ITEM		
SUB-TOTAL		
OTHER DEDUCT. (IF ANY)		
BAL. FWD.		

Ullman & Company
390 Cherry Rd.
Boise, Idaho 54948

3089

63-398/670

______ 19 ____

PAY TO THE ORDER OF ______ $ ______

______ DOLLARS

Barnett Bank
037-049
11755 Biscayne Blvd.
North Miami, Florida 33161

FOR ______

⑆067003985⑆ 3089 630211502 ⑈

Ullman & Company
390 Cherry Rd.
Boise, Idaho 54948

DATE ______ 19 ____

DEPOSITS MAY NOT BE AVAILABLE FOR IMMEDIATE WITHDRAWAL

______

SIGN HERE IF CASH RECEIVED FROM DEPOSIT

CASH: CURRENCY		
CASH: COIN		
CHECKS		
TOTAL FROM OTHER SIDE		
TOTAL		
LESS CASH		
NET DEPOSIT		

63-398/670

DEPOSIT TICKET

USE OTHER SIDE FOR ADDITIONAL LISTINGS

TOTAL ITEMS ☐

BE SURE EACH ITEM IS PROPERLY ENDORSED

Barnett Bank
037
11755 Biscayne Blvd.
North Miami, Florida 33161

⑆067003985⑆ 630211502 ⑈

REV. 6/88 CHECKS AND OTHER ITEMS ARE RECEIVED FOR DEPOSIT SUBJECT TO THE PROVISIONS OF THE UNIFORM COMMERCIAL CODE OR ANY APPLICABLE COLLECTION AGREEMENT

2. You have just received a check. Your account number is #265-55-7. Write the following endorsements in the space provided below, and identify what type they are:
   a. Allowing you to deposit the check into your account.
   b. Allowing you to cash the check.
   c. Allowing the check to be transferred to Beverly Baker.

a. ____________________ b. ____________________ c. ____________________

____________________ ____________________ ____________________

____________________ ____________________ ____________________

____________________ ____________________ ____________________

3. On October 1, Geoff Seeger received his bank statement showing a balance of $788.28. His checkbook records indicate a balance of $852.80. There was a service charge for the month of $9.60 on the statement. The outstanding checks were for $42.50, $12.80, $108.17, and $27.72. The deposits in transit totaled $246.11. Use the form below to reconcile Geoff's checking account.

CHECKBOOK BALANCE	$	STATEMENT BALANCE	$
Deduct: Service Charges & Other Debits		Deduct: Outstanding Checks	
SUB TOTAL		SUB TOTAL	
Add: Interest Earned & Other Credits		Add: Deposits in Transit	
ADJUSTED CHECKBOOK BALANCE		ADJUSTED STATEMENT BALANCE	

Reconciled Balances

Checks Outstanding

No.	Amount	
Total		

4. On September 5, Ann Rayner received her bank statement showing a balance of $5,595.37. Her checkbook showed a balance of $5,489.01. The statement showed that the bank had collected a note for Ann in the amount of $2,000.00. There was also a returned item of $550.00, and a nonsufficient funds (NSF) charge of $25.00. The deposits in transit totaled $3,261.14, and the outstanding checks were for $262.07, $467.50, and $1,212.93. Use the form below to reconcile Ann's account.

CHECKBOOK BALANCE	$	STATEMENT BALANCE	$
Deduct: Service Charges & Other Debits		Deduct: Outstanding Checks	
SUB TOTAL		SUB TOTAL	
Add: Interest Earned & Other Credits		Add: Deposits in Transit	
ADJUSTED CHECKBOOK BALANCE		ADJUSTED STATEMENT BALANCE	

Reconciled Balances

Checks Outstanding

No.	Amount	
Total		

5. You are the accountant for Triple Crown Jewelers. On September 4, you receive the following statement from Grove Isle Bank. Using the form below, reconcile the statement with the checkbook records.

Grove Isle Bank

STATEMENT DATE
9-2-19xx

Triple Crown Jewelers
437 Ventura Blvd.
Louisville, Ky. 40202

CHECKING ACCOUNT SUMMARY
8-1-19xx THRU 8-31-19xx

ACCOUNT NUMBER
8004-393-22

Previous Balance	Deposits & Credits Number	Total	Checks & Debits Number	Total	Current Balance
1,237.75	4	1,331.25	7	1,462.94	1,106.06

CHECKING ACCOUNT TRANSACTIONS

DATE	AMOUNT	DESCRIPTION	BALANCE
8-2	246.12	Check #0501	991.63
8-4	350.73	Deposit	1,342.36
8-11	125.00	Returned Item	1,217.36
8-13	454.90	EFT Payroll Deposit	1,672.26
8-14	275.69	Check #0503	1,396.57
8-17	493.88	Check #0505	902.69
8-23	175.00	ATM Withdrawal	727.69
8-24	225.62	Deposit	953.31
8-27	129.25	Check #0506	824.06
8-30	18.00	Service Charge	806.06
8-31	300.00	Note Collected	1,106.06

PLEASE BE SURE TO **DEDUCT** ANY BANK CHARGES THAT APPLY TO YOUR ACCOUNT.

CHECK NUMBER	DATE	DESCRIPTION OF TRANSACTION	AMOUNT OF PAYMENT OR WITHDRAWAL (−)		✓	AMOUNT OF DEPOSIT OR INTEREST (+)		BALANCE FORWARD	
								1,237	75
0501	8/1	To First Credit Corp.	246	12				991	63
		For						Bal.	
0502	8/3	To T.J.G. Supplies, Inc.	319	43				672	20
		For						Bal.	
	8/4	To Deposit				350	73	1,022	93
		For						Bal.	
	8/11	To Deposit				454	90	1,477	83
		For						Bal.	
0503	8/11	To King Advertising	275	69				1,202	14
		For						Bal.	
0504	8/15	To Grove Supply Co.	186	17				1,015	97
		For						Bal.	
0505	8/17	To Southern Gold, Inc.	493	88				522	09
		For						Bal.	
	8/23	To ATM Withdrawal	175	00				347	09
		For						Bal.	
	8/23	To Deposit				225	62	572	71
		For						Bal.	
0506	8/25	To Technical Tools Co.	129	25				443	46
		For						Bal.	
	8/31	To Deposit				100	00	543	46
		For						Bal.	
		To							
		For						Bal.	
		To							
		For						Bal.	

Checks Outstanding

No.	Amount	
Total		

CHECKBOOK BALANCE	**$**	**STATEMENT BALANCE**	**$**
Deduct: Service Charges & Other Debits		**Deduct:** Outstanding Checks	
SUB TOTAL		**SUB TOTAL**	
Add: Interest Earned & Other Credits		**Add:** Deposits in Transit	
ADJUSTED CHECKBOOK BALANCE		**ADJUSTED STATEMENT BALANCE**	

Reconciled Balances

Set up and solve equations for the following business situations:

1. Tony and George sell cars for Classic Motors. Last month, George sold 6 less cars than Tony. If they sold a total of 46 cars, how many did each sell?

2. One-sixth of the trucks at Northwest Transport Corporation are Fords. If 540 of the trucks are Fords, what is the total number of trucks in the Northwest fleet?

3. All of the 244 stockholders attended the annual meeting of InterAmerican Industries, Inc. If the total cost of the function was $6,222, what was the cost per stockholder?

4. A shoe store sells $3\frac{1}{2}$ times as much in women's shoes as in men's shoes. If last month's sales totaled $65,430, how much of each type of shoe was sold?

5. The San Francisco plant of the AirJet Corporation can produce 200 more than twice the production per hour of the Los Angeles plant. If the San Francisco plant produces 1,500 units per hour, how many does the Los Angeles plant produce?

6. A gift shop ordered picture frames from a distributor. Large frames cost $7.50 each and small frames cost $4.50 each.

a. If the total order was for 88 frames and amounted to $471, how many of each size were ordered?

b. What was the dollar amount of each size?

7. Lucas and Brown invested $128,000 in a grocery store joint venture. If Brown invested four times as much as Lucas, how much did each invest?

8. A lottery is to be distributed among four winners. Jim and Fred each had one ticket, Lisa had three tickets, and Annie had four tickets. If the total lottery prize was $376,200, how much should each winner receive?

9. Last week, Office Depot found that 14 less than one-fourth of its sales of calculators was for solar-powered models. If 44 solar-powered calculators were sold, what was the total number of calculators sold?

10. Cellular telephones cost $35 more than three times the cost of cordless phones. If together they cost $549, what is the cost of each type of phone?

11. House Beautiful, a furniture store, sells leather sofas for $900 and fabric sofas for $650.

a. If 36 sofas were purchased from the manufacturer for a total of $28,400, how many of each type were purchased?

b. What is the dollar value of each type sofa?

12. The Winston building is 3 times as high as the Blakley Building. If the difference between the two is 88 feet, what is the height of each building?

13. One-twentieth of a company's employees have over twenty years of service. If 42 employees are in that category, how many people work for the company?

14. The Fancy Fish Market sells shrimp for \$8.50 per pound and scallops for \$6.50 per pound. They sell twice as many pounds of shrimp as scallops. If total sales yesterday amounted to \$1,316, how many pounds of each were sold?

15. A set of dishes, which serves eight people, contains five pieces per place setting. If the total price of the set is \$480, what is the price per piece?

16. What is the total cost to ship an order weighing 350 pounds if the breakdown is \$.15 per pound for packing, \$.08 per pound for insurance, \$.27 per pound for transportation, and \$42.50 for the crate?

Use ratio and proportion to solve the following business situations:

17. If the interest on a \$2,500 loan is \$278, what would be the interest on a loan of \$3,100?

18. At FunFair, the ratio of hamburgers to hot dogs sold is 7 to 4. If 350 hamburgers were sold on a busy weekend, how many hot dogs were sold?

19. If disability insurance costs \$1.25 per \$100 of coverage, how much insurance would be available for a premium of \$70.00?

20. A recipe calls for $\frac{1}{2}$ teaspoon of vanilla for every $1\frac{1}{2}$ cups of fresh fruit. If a large fruit salad requires 9 cups of fruit, how much vanilla should be used?

21. A map of the city uses a scale of $\frac{1}{4}$ inch equals 5.5 miles. What is the distance in miles of a trip that measures 8 inches on the map?

22. If a car can travel 330 miles on 15 gallons of fuel, how far can it go on 55 gallons?

23. If a knitting machine can make 16 sweaters every hour and a half, how many sweaters can it make in 22.5 hours?

24. An amusement park is visited by people in the ratio of seven children for every three adults. If 450 adults attended in one day, how many children attended?

Solve the following word problems for the unknown, rounding decimals to hundredths and percents to tenths:

1. Last season, the Miami Dolphins won 12 and lost 5 games. What percent of the games did they win?

2. John Lannefeld earned $83,500 last year. If he got a 5% raise, what is his new salary?

3. At Circuit City, the price of a stereo went from $660 to $548. What is the rate of decrease?

4. In a recent survey, 306 people preferred instant ice tea over fresh brewed. If that represented 18% of the total respondents, how many people were interviewed?

5. The Ronson Corporation employs 860 people in three regions of the country. If 30% are in the Midwest region, and 50% are in the Northeast region, how many people are employed in the Southern region?

6. If the sales tax on a purchase of $4,700 amounts to $387.75, what is the tax rate?

7. A toy manufacturer gets 34% of its revenue from doll houses. If doll house sales last year were $527,000, what were the total sales of the company?

8. A business is owned by three partners as follows: Ethel owns 27%, Goldie owns 37%, and Fred owns the rest. If Fred's share amounts to $45,000, what is the dollar amount of Ethel's and Goldie's shares?

9. Wes Lawton sells appliances for Apple Distributors. Last week he earned $990 in commissions. If his commission rate is $8\frac{1}{4}$%, what was the total amount of his sales?

10. A trade show had 564 exhibitors last year and 768 this year. What is the percent increase in exhibitors?

11. A plant nursery sold $2,750 in orchids in August. If September sales were 18% less, what were the orchid sales in September?

12. Mary Jo Kovach is paid a gross salary each week of $564.00. After deductions for taxes, social security, and Medicare, she receives a check for $434.28.

a. What percent of her income is taken out for deductions?

b. If Mary Jo got a raise to $31,200 per year, based on the same percent of deductions, what would be her new weekly take-home pay?

13. Colgate Toothpaste had a 32% share of the market last year, and a 37% share this year. If this represents an increase of 5 percentage points, what is the percent increase in Colgate Toothpaste sales?

14. A company depreciates its new trucks and equipment for five years, and then sells them to used equipment dealers.

a. If the company paid $8,900 for a forklift five years ago and sold it for $1,335 last week, what percent of the original price was depreciation?

b. Using the same rate of depreciation as with the forklift, if the company sold a conveyor belt transport system to a used equipment dealer for $1,848, what was the original purchase price of this equipment?

15. Ocean City has a population today of 665,000 people.

a. If this represents $17\frac{1}{2}\%$ of the state's population, how many people live in the state?

b. If the city's population today is 250% of what it was ten years ago, how many people lived there ten years ago?

c. Research indicates that in five years the population of the city will be 20.2% higher than today's population. Based on this prediction, how many people will be living there in five years?

16. An ice cream vendor pays $17.50 for a five-gallon container of premium ice cream. From this quantity, he sells 80 scoops at $.90 per scoop. If he sold smaller scoops, he could sell 98 scoops from the same container, however he could only charge $.80 per scoop. As his accountant, you are asked the following questions:

a. If he switches to the smaller scoops, by how much will his profit per container go up or down? (Profit = Sales – Expenses.)

b. By what percent will the profit change?

17. At the end of the day, a fruit stand had total receipts of $1,300 in the cash register. If this included revenue from the fruit sales and 4% state sales tax collected, how much of the total receipts were from fruit sales?

1. King's Dry Cleaners receives an invoice for the purchase of chemicals with a list price of $1,500. Because they are in the trade they get a 25% trade discount. What is the amount of the trade discount?

2. The Computer Connection buys a computer listing for $2,620, less a 30% trade discount.

   a. What is the amount of the trade discount?

   b. What is the net price of the computer?

3. Use the net price factor method (complement method) to verify your net price answer in Question 2.

4. The Cuisine Supreme Restaurant places an order with a food supplier listing for $2,230. They receive a trade discount of $528 on the order. What is the trade discount rate? (Round to the nearest hundredth percent.)

5. The Art Emporium can purchase drawing pads from Premium Paper Company for $50 per dozen less trade discounts of 15/10/7. A competitor, Econo Paper Company, is offering the same item for $55 less trade discounts of 18/10/6. Which offers a lower net price?

6. Tool Town purchased the following items. Calculate the extended total for each line, the invoice subtotal, and the invoice total.

Quantity	Unit	Merchandise	Unit List	Discounts	Extended Total
250	ea.	3/8" Drills	$ 89.95	20/15/10	______
400	ea.	Belt Sanders	69.88	25/10/7	______
124	ctn.	Sharpeners	274.39	30/14/7	______
				Invoice subtotal	______
				Extra $4\frac{1}{2}$% volume discount	______
				Invoice total	______

7. What are the net price and the trade discount amount on merchandise with a list price of $10,000 less trade discounts of 25/20/8?

8. What is the net price factor for trade discounts of 20/15/12? Use that net price factor to find the net price of a refrigerator listing for $1,300.

9. What are the net price factor and the single equivalent discount of the trade discount series 30/15/5?

10. What are the net price factor and the single equivalent discount of the trade discount series 25/10/5?

11. Explain the following terms of sale:

a. 3/15, n/30

b. 2/20, n/60

c. 4/20, 2/30, n/60

d. 3/15, n/45 ROG

e. 4/15 EOM

f. 2/15, 40 Extra

12. Beyond Electronics receives an invoice for an order of video games dated January 15, with terms of 3/15, n/30.

a. What is the discount date? What is the net date?

b. If the invoice terms are changed to 3/15, 20 Extra, what is the discount date and what is the net date?

13. Lynn's Gift Shop buys pottery from a supplier for $2,750 with terms of 4/20, n/45. The invoice is dated June 28. Lynn decides to send in a $1,200 partial payment within the discount period.

a. What are the discount date and the net date?

b. What is the amount of credit Lynn will receive for the partial payment?

c. What is the balance still due the supplier?

Solve the following word problems, rounding dollars to the nearest cent and percents to the nearest tenth:

1. Calculators cost the manufacturer $33.00 per unit to produce. If a markup of $9.80 is added to the cost, what is the selling price per calculator?

2. A department store sells stockings for $6.88 per pair. If the stockings cost $2.76 per pair, what is the amount of the markup?

3. After Royal Beauty Supply adds a markup of $25.30 to a hair dryer, it is sold for $73.00. What is the cost of the hair dryer?

4. Computer Village buys printers for $435.60 and sells them for $675.90.

   a. What is the amount of the markup?

   b. What is the percent markup based on cost?

5. Handy Hardware purchases hammers for $6.70. A $4.75 markup is added.

   a. What is the selling price?

   b. What is the percent markup based on cost?

6. A picture frame has a cost of $19.95 and a selling price of $31.50.

   a. What is the amount of the markup?

   b. What is the percent markup based on selling price?

   c. What is the percent markup based on cost?

7. Blue Water Marine purchases boats at a cost of $26,570 and sells them for $48,295.

   a. What is the amount of the markup?

   b. What is the percent markup based on selling price?

   c. What is the percent markup based on cost?

8. Appliance Express purchases fans from the manufacturer for $72.00. If the store policy is to mark up all merchandise in that department 46% based on selling price, what is the retail selling price of the fans?

9. A buyer for a women's apparel retailer is looking for a line of bathing suits to sell for $139.95. If the objective is a 58% markup based on selling price, what is the most that the buyer can pay for the suits to still get the desired markup?

10. a. If the markup on a stereo is 27% based on selling price, what is the corresponding percent markup based on cost?

b. If the markup on a diamond ring is 300% based on cost, what is the corresponding percent markup based on selling price?

11. A wallet that originally sold for $45.50 was marked down and sold for $29.95.

a. What is the amount of the markdown?

b. What is the markdown percent?

12. An airline ticket from New York to San Francisco that originally sold for $439.00 was marked down by $88.00.

a. What is the sale price of the ticket?

b. What is the markdown percent?

13. The Pampered Pet originally sold bird cages for $53.50. If they are put on sale at a markdown of 30%, what is the sale price?

14. What was the original selling price of a weight bench, currently on sale at Home Fitness Studio for $224.66 after a 25% markdown?

15. In July, a men's shop purchased sport coats for $124.50 each. The original markup was 48% based on selling price. In September, the shop took a 25% markdown by having a special sale. After three weeks, the sale was over and prices were marked up 18%. In November, it offered a storewide holiday sale of 20% off all merchandise, and in January, a final 15% markdown was taken. What was the final selling price of the sport coats?

16. The Fruit Nut purchases 250 pounds of strawberries at $.69 per pound. If a 20% spoilage rate is anticipated, at what price per pound should the strawberries be sold in order to achieve a 45% markup based on cost?

17. The weight bench originally sold for $279.96 and the weights sold for $55.96.

a. What is the amount of markdown for each?

b. What is the markdown percent of each?

18. What was the original selling price of the New Balance running shoes?

19. **a.** What is the cost to the retailer of the Huffy bicycle if the markup is 35% based on selling price?

b. If the store offered a 20% markdown from the $199.96 price, what would be the new sale price?

1. Mickey Brittain has gross earnings of $2,650 biweekly. If his company switches pay schedules from biweekly to semimonthly, what will Mickey's new gross earnings be?

2. Kanny Washburn works 40 hours per week. At the rate of $9.45 per hour, what are Kanny's gross weekly earnings?

3. Eilish Main works for a company that pays time-and-a-half for all hours worked over 40. If Eilish earns $12.50 per hour, what are her gross weekly earnings for a 55-hour week?

4. Roger MacQuarrie earns $6.30 per hour for regular time up to 40 hours, time-and-a-half for overtime, and double time for the midnight shift. Last week Roger worked 61 hours, including 7 on the midnight shift. What are his gross earnings?

As payroll manager for Twin Lakes Manufacturing, it is your task to complete the following weekly payroll record. The company pays overtime for all hours worked over 40 at the rate of time-and-a-half:

	Employee	M	T	W	T	F	S	S	Hourly Rate	Total Hours	Overtime Hours	Regular Pay	Overtime Pay	Total Pay
5.	Proctor	6	8	7	8	9	0	0	$6.75	______	______	______	______	______
6.	Wilson	7	9	12	8	11	0	6	$9.00	______	______	______	______	______
7.	Chalks	5	8	7	8	9	0	7	$13.70	______	______	______	______	______

8. Randy Terry assembles motors for Eastern Electric. He is paid on a straight piecework rate of $4.19 per motor. If he assembled 108 units last week, what were his weekly gross earnings?

9. Bill Clements works for Contempo Casuals, a company that manufactures men's clothing. Bill is paid $2.30 for each pair of slacks, $4.15 for each suit, and $3.65 for each jacket. If Bill produced 46 slacks, 21 suits, and 36 jackets, what were his total weekly gross earnings?

You are the payroll manager for Maxwell Electronics, a manufacturer of cellular phones and beepers. Your workers are paid on a differential piecework schedule as follows:

Pay Level	Units Produced	Rate per Unit
1	1–75	\$3.12
2	76–140	\$3.45
3	141–200	\$4.08
4	Over 200	\$4.40

Calculate the amount of last week's total gross pay for the following employees:

	Employee	Units Produced	Total Gross Pay
10.	Peterson, B.	68	______
11.	Jiminez, L.	128	______
12.	Morgan, S.	233	______

13. What is the total gross pay for a salesperson on a straight commission of 5.1% if her sales volume is \$153,000?

14. What is the total gross pay for a salesperson on an incremental commission schedule of 1.8% of sales up to \$220,000 and 2.3% on sales over \$220,000, if last month's sales volume was \$385,300?

15. Todd McGarity is a salesperson for a company that pays a salary of \$575 per month plus a commission of $6\frac{1}{2}$% of all sales over \$12,000. If Todd sold \$35,300 last month, what was his total gross pay?

16. John Belliveau is on a 3.6% straight commission with a \$1,400 drawing account. If he is paid the draw at the beginning of the month and then sells \$172,000 during the month, how much commission is owed to John?

17. Sally Swift, a waitress at the Chart House Restaurant, is paid a salary of \$24.50 per day. She also averages tips of 17% of her total gross food orders. Last week Sally worked 5 days and had total food orders of \$1,840. What were her total gross pay and tips for the week?

18. David Sporn is a real estate broker for International Property Sales, Inc. The company makes 6% commission on all sales. David's share is 45% of what the company takes in on his sales. Last week David closed a sale on a penthouse apartment at Yacht Harbour for $650,000.

a. How much did International take in on the deal?

b. How much was David's share of the commission?

Solve the following problems using 6.2% up to $62,700 for social security, and 1.45% for Medicare:

19. What are the withholdings for social security and Medicare for an employee with gross earnings of $725 per week?

20. What are the withholdings for social security and Medicare for an employee with gross earnings of $1,663 paid semimonthly?

21. Anita Louise is an executive with the Kimberly Corporation. Her gross earnings are $8,800 per month.

a. What are the withholdings for social security and Medicare for Anita's January paycheck?

b. In what month will Anita's salary reach the social security wage base limit?

c. What are the social security and Medicare tax withholdings for Anita in the month named in part b?

As payroll clerk for Sweetheart Bakeries, it is your task to calculate the monthly social security and Medicare withholdings for the following employees:

Employee	Year-to-Date Earnings	Current Month	Social Security	Medicare
22. Kloot, L.	$24,551	$2,900	______	______
23. Jefferson, G.	$59,810	$4,100	______	______
24. Rodriguez, J.	$155,090	$11,150	______	______

Use Exhibit 9-2 and Exhibit 9-3 to solve the following problems:

25. How much income tax should be withheld from the paycheck of a married employee earning $1,245 per week and claiming five withholding allowances?

26. How much income tax should be withheld from the paycheck of a single employee earning $3,250 per month and claiming one withholding allowance?

27. Lisa Miller is single, claims one withholding allowance, and earns $2,620 per month. What is the amount of Lisa's paycheck after her employer withholds social security, Medicare, and federal income tax?

Use the *wage bracket method* of income tax calculation to complete the following payroll roster:

Employee	Marital Status	Withholding Allowances	Pay Period	Gross Earnings	Income Tax Withholding
28. Alders, M.	S	2	Monthly	$4,346	______
29. Nunez, C.	M	4	Weekly	$921	______
30. Burkett, S.	M	6	Weekly	$1,218	______
31. Wong, A.	S	3	Monthly	$3,065	______

Use the *percentage method* of income tax calculation to complete the following payroll roster:

Employee	Marital Status	Withholding Allowances	Pay Period	Gross Earnings	Income Tax Withholding
32. Rockwell, V.	M	4	Weekly	$688.00	______
33. Kelman, P.	S	0	Semimonthly	$1,268.00	______
34. Scher, A.	S	2	Monthly	$5,210.50	______
35. Segall, S.	M	3	Biweekly	$1,702.00	______

36. William Price is married, claims five withholding allowances, and earns $5,010 per month. In addition to social security, Medicare, and FIT, William also pays 2% state income tax, $\frac{1}{2}$% for state disability insurance (both based on gross earnings), $33.60 for term life insurance, and $112 to his credit union for a boat payment. As payroll manager for William's company, calculate his net take-home pay per month.

37. Mayfield Manufacturing has 43 employees on the assembly line, each with gross earnings of $428 per week.

a. What is the total social security and Medicare tax that should be withheld from the employee paychecks each week?

b. What is Mayfield's share of the social security and Medicare tax for the first quarter of the year?

38. Brian Wilson is a self-employed carpenter. Last year he had total gross earnings of $24,960. How much does Brian have to send the IRS each quarter for social security and Medicare?

As the accounts payable manager for New Age Associates, complete the following payroll record using the percentage method for tax calculation:

	Employee	YTD Earnings	Pay Period	This Period	Marital Status	Allow. Claimed	Social Security	Medicare	Income Tax	Other	Net Pay
39.	Mann	$19,300	Weekly	$516	S	0	______	______	______	$88	______
40.	Tolbert	27,540	Biweekly	3,762	M	3	______	______	______	47	______
41.	Eason	59,100	Semimonthly	3,933	S	2	______	______	______	174	______
42.	Best	195,690	Monthly	15,200	M	6	______	______	______	264	______

43. Liz Widdicombe earns $55,400 annually as a supervisor for Turnbuckle Engineering.

a. If the SUTA tax rate is 5.4% of the first $7,000 earned in a year, how much SUTA tax must Turnbuckle pay each year for Liz?

b. If the FUTA tax rate is 6.2% of the first $7,000 earned in a year minus the SUTA tax paid, how much FUTA tax must the company pay each year for Liz?

44. Bugs-Away Exterminators has three employees. Tinkers earns $350 per week, Evers earns $470 per week, and Chance earns $735 per week. The company's SUTA tax rate is 5.4%, and the FUTA rate is 6.2% minus the SUTA. As usual, these taxes are paid on the first $7,000 of each employee's earnings.

a. How much SUTA and FUTA tax does the company owe on these employees for the first quarter of the year?

b. How much SUTA and FUTA tax does Bugs-Away owe for the second quarter of the year?

45. Allied Enterprises employs 153 workers and has a gross payroll of $254,000 per week. Benefits are 5% of gross payroll for sick days and holiday leave, 7.1% for health and hospital insurance, 3.6% for the retirement fund, and $16.50 per employee for a stock purchase plan.

a. What is the total weekly cost of benefits for the company?

b. What percent of payroll does this represent? (Round to the nearest tenth percent.)

c. What is the company's annual cost of benefits?

46. Stacey Sims is self-employed with an estimated annual salary of $56,600. Her social security tax rate is 12.4% and her Medicare is 2.9%. This year Stacey estimates that her federal income tax rate will be 17%.

a. How much quarterly estimate tax must Stacey send to the IRS each quarter?

b. What form should she use?

47. You are the payroll manager for Electronic Wholesalers. Last year your company's tax liability was $41,000.

a. According to IRS rules, how often should the tax deposit be made?

b. What form should accompany the deposit?

c. How often must you file an employer's tax return for the company?

d. What IRS form should you use for this purpose?

1. On September 2, Candace Cooney borrowed $800 from her credit union at 11%, for 120 days. The credit union uses the ordinary interest method.

   a. What is the amount of interest on the loan?

   b. What is the maturity value of the loan?

   c. What is the maturity date of the loan?

2. The Downtown Appliance Company borrowed $440,000 at $12\frac{1}{2}$% interest on April 30, for 73 days.

   a. If the bank uses the exact interest method, what is the amount of interest on the loan?

   b. What is the maturity date?

3. Lionel Howard missed an income tax payment of $7,800. The Internal Revenue Service charges a 14% simple interest penalty, calculated by the exact interest method. If the tax was due on April 15 but was paid on June 7, what is the amount of the penalty charge?

4. Lee Monroe borrowed money to buy a car from her credit union at 12.7% simple interest. If the loan was repaid in 3 years and the amount of interest was $4,572, how much did Lee borrow?

5. Joe Moutran took out a loan for $5,100 at the Northern Trust Bank for 240 days. The bank uses the ordinary method for calculating interest. What rate of interest was charged if the amount of interest was $335? Round to the nearest tenth percent.

6. Alumicraft, a manufacturer of truck bodies, borrowed $1,355,000 on November 12 to purchase a shipment of aluminum. The interest rate was 11.8% using the ordinary interest method. The amount of interest was $46,200.

   a. For how many days was the loan? Round any portion of a day to the next higher day.

   b. What was the maturity date of the loan?

7. Elizabeth Hayes deposited $4,000 in a savings account paying 6.8% ordinary interest. How long will it take for her investment to amount to $5,360?

8. Dona Hightower borrowed $12,000 at 13% ordinary interest, for 70 days. On day 25 of the loan she made a partial payment of $5,000. What is the new maturity value of the loan?

9. Hi-Volt Electrical Supplies borrowed $80,000 on May 26 for 80 days. The rate was 14% using the ordinary interest method. On day 40 of the loan Hi-Volt made a partial payment of $21,000, and on day 65 of the loan Hi-Volt made a second partial payment of $15,000.

   a. What is the new maturity value of the loan?

   b. What is the maturity date of the loan?

10. Ralph Waldo signed a $35,000 simple discount promissory note at the First Union Bank. The discount rate was 13% ordinary interest, and the note was made on June 19, for 70 days.

   a. What proceeds will Ralph receive on the note?

   b. What is the maturity date of the note?

11. Edgar Allan signed a $15,000 simple discount promissory note at a bank discount rate of 12.4%. If the term of the note was 115 days, what is the effective rate of interest? Round your answer to the nearest hundredth percent.

12. Boston Whaler Boat Manufacturing received a $280,000 promissory note at 13% simple interest for 85 days from one of its dealers. After 60 days Boston Whaler discounted the note at Citibank at a discount rate of 16%. The note was made on February 11.

   a. What was the maturity date of the note?

   b. What was the maturity value of the note?

   c. What was the discount date of the note?

   d. What proceeds did Boston Whaler receive after discounting the note?

CHAPTER 11 PRACTICE EXERCISES

Solve the following word problems using either Tables 11-1 and 11-2 or the optional compound interest and present value formulas. When necessary, create new table factors (round dollars to the nearest cent and percents to the nearest hundredth):

1. Lois Pahls invested $30,000 at 10% interest compounded semiannually at the Great Lakes Bank. How much will Lois's investment be worth after 6 years?

2. Barry Schwartz will be graduating from medical school in 33 months. If he invests $125,000 now at 8% interest compounded quarterly, will he earn enough interest to have the $150,000 down payment needed to set up his medical practice?

3. David Chwalik wants $60,000 in 10 years as a college fund for his 8-year-old son. How much must David invest today at 12% interest compounded semiannually, in order to have the money when needed?

4. Debbie Bacus wants $5,000 in 4 years for a trip to Australia. If her bank is paying 6% interest compounded monthly, how much should Debbie invest now in order to have enough for the trip?

5. What is the effective rate of a $10,000 investment, for one year, at 12% interest compounded monthly?

6. Diane Southworth invested $25,000 for 5 years at 10% interest compounded semiannually.

   a. What is the compound amount?

b. What is the amount of compound interest?

c. What is the effective rate?

7. Bill Schoof wants to buy a Winnebago motor home in 5 years to tour the United States. How much must Bill invest today at 8% compounded quarterly in order to have the $37,500 needed for the vehicle?

8. Nancy Weaver invested $7,000 at the Southwestern Bank at 8% interest compounded quarterly.

a. What is the effective rate of this investment?

b. What will Nancy's investment be worth after 10 years?

c. How much more would Nancy's investment be worth after 10 years if she could have gotten 10% interest compounded semiannually, instead of 8% quarterly?

9. Bob DeLucia and his family are planning a vacation to Alaska in $3\frac{1}{2}$ years. Bob estimates that he will need $6,800 for the trip.

a. If The Big East Bank is offering a $3\frac{1}{2}$-year certificate of deposit with 8% interest compounded quarterly, how much should Bob invest now in order to have the money for the trip?

b. How much compound interest will be earned on the investment?

10. Dan and Myra Kahlani deposited $15,000 in an account at the Oahu State Bank paying 6% interest compounded quarterly, when their daughter was born. When she was 5 years old they added $10,000 to the account, which was now paying 8% compounded semiannually. How much will the account be worth when their daughter is 17 years old?

11. Over the past 10 years Steve Nadler had the following investment experience:

1. Deposited $10,000 at 8% compounded semiannually, in a 3-year certificate of deposit.
2. After 3 years, he cashed in that CD and added another $5,000, to buy a 4-year, 6% certificate compounded quarterly.
3. When that certificate matured, Steve added another $8,000 and bought a 3-year, 7% certificate compounded annually.

a. What was the total worth of his investment when the last certificate matured?

b. What is the total amount of compound interest earned over the 10-year period?

12. Southside Savings is offering certificates of deposit at 6% interest compounded quarterly; Northside National is offering certificates at 7% compounded annually. If you had invested $10,000 for 5 years, how much interest would have been lost by choosing the wrong bank?

1. How much will $1,000 deposited at the *beginning* of each 6-month period be worth after 5 years at 6% interest compounded semiannually?

2. How much must be deposited now in order to withdraw $500 at the *end* of each month for one year, if interest is 18% compounded monthly?

3. How much must be deposited now in order to withdraw $1,200 at the *beginning* of each year for 15 years, if interest is 9% compounded annually?

4. Russ Boersma is planning for his retirement. He decides to deposit $1,000 at the beginning of each 3-month period into an account paying 8% interest compounded quarterly.

   a. How much will Russ have in the account at the end of 9 years?

   b. What is the total amount of interest earned on the account?

   c. How much would the account be worth after 9 years if Russ decides to deposit $2,000 at the beginning of each period instead of $1,000?

5. A professional basketball team has signed a player whose contract calls for a $100,000 bonus at the end of each year, for the next 5 years. How much should the team deposit now at 9% interest compounded annually, in order to meet this financial obligation?

6. The Manners family is planning a European vacation in $3\frac{1}{2}$ years and estimates they will need \$6,000 for the trip. They set up a sinking fund savings account whereby regular payments are made into the account at the end of each 6-month period.

   a. If the interest is 6% compounded semiannually, what periodic sinking fund payment is required for the Manners to reach their goal?

   b. What payment would be required if they were able to find a bank paying 8% compounded semiannually, rather than 6%?

7. Jeff Ehmke bought a living room set for \$3,500 at Contempo Furniture. He made a \$500 down payment and is financing the balance at the store over a 2-year period at 18% interest.

   a. What equal monthly payments will be required by Jeff to amortize the loan?

   b. If Jeff's credit union will finance the furniture for 12% compounded monthly, how much can he save per month on the payments?

8. Thai Industries has a \$2,000,000 financial obligation due in 6 years. The company establishes a sinking fund at 10% interest compounded semiannually.

   a. What equal payments must be deposited by the company every 6 months in order to meet this financial obligation?

   b. What is the total amount of interest earned in this sinking fund account?

9. Ralph Kramden has been depositing \$300 at the beginning of each month into a savings account for 2 years. If his bank has been paying 6% interest compounded monthly, how much does Ralph have in the account?

10. Comstock Enterprises has been awarded a settlement in a lawsuit of $10,000 at the end of each year for 5 years. How much must be set aside now at 7% interest compounded annually in order to pay this obligation to Comstock?

11. The Millennium Corporation purchased an airplane for $560,000. The company made a 10% down payment and financed the balance for 3 years at 12% interest.

a. What equal payments must be made every 3 months by Millennium to amortize this loan?

b. How much would the payments be if the loan was extended to 4 years?

(Optional) Use formulas and a financial calculator to solve the following word problems:

12. Nightingale Bank is paying 6% interest compounded monthly. Use the formula for an ordinary annuity to calculate how much $500 deposited at the end of each month would amount to in 5 years.

13. Morningstar Savings is paying 8% interest compounded semiannually. Use the formula for an annuity due to calculate how much $1,000 deposited at the beginning of each 6-month period would amount to in 25 years.

14. Use the formula for the present value of an ordinary annuity to calculate the amount that must be deposited now at 10% interest compounded quarterly, to yield an annuity payment of $1,000 at the end of each 3-month period for 10 years.

15. Use the formula for the present value of an annuity due to calculate the amount that must be deposited now at 6% interest compounded monthly, to yield an annuity payment of $100 at the beginning of each month for 5 years.

16. Record World, a chain of music stores, is planning to expand in 5 years. It is estimated that $3,000,000 will be needed for the project. If the company sets up a sinking fund to save for this obligation, what equal monthly payments must be deposited into the account at 9% compounded monthly to achieve the goal?

17. Larry Schweiger is interested in buying a yacht for $87,000. His bank is willing to finance the purchase at 12% interest for 8 years. The bank also requires a 20% down payment. What equal monthly payments will be required by Larry to pay off this loan?

1. John Anderson has a BankCard account with an annual percentage rate of 15%, calculated on the previous month's balance. During November John's account showed the following activity.

John Anderson **STATEMENT OF ACCOUNT**

Account #5490-0890-4939 **Billing cycle: November 1–30**

November 1	Previous Month's Balance	$461.00
November 6	Montgomery Ward	88.67
November 7	Payment	55.00
November 17	Shell Oil	19.79
November 26	Cash Advance	100.00
November 27	Innovations Gift Shop	121.40

a. What is the amount of the finance charge?

b. What is John's new balance?

Penny Mager has a revolving charge account. The finance charge is calculated on the previous month's balance, and the annual percentage rate is 18%. Complete the following 4-month account activity table for Penny:

Month	Previous Month's Balance	Finance Charge	Purchases and Cash Advances	Payments and Credits	New Balance End of Month
2. August	$740.00	______	$218.66	$125.00	______
3. September	______	______	$56.48	$75.00	______
4. October	______	______	$185.35	$85.00	______
5. November	______	______	$322.65	$270.00	______

6. Calculate the average daily balance for the month of May of an account with a previous month's balance of $530.24 and the following activity:

Date	Activity	Amount
May 2	Cash Advance	$150.00
May 8	Payment	$135.00
May 13	Purchase	$59.61
May 19	Credit	$35.34
May 26	Purchase	$202.78

7. Eddie Rhodes is a salesperson for Pan Am Enterprises. When he travels, Eddie uses his BankCard to pay for many of his business expenses. The account has a 14% annual percentage rate calculated on the average daily balance. The billing date is the first day of each month and the billing cycle is the number of days in that month. During the month of October Eddie's account showed the following activity:

Eddie Rhodes	STATEMENT OF ACCOUNT	
Account #4495-5607	Billing cycle: October 1–31	
October 1	Previous month's balance	$326.77
October 3	Regency Hotel	454.30
October 7	Payment	500.00
October 14	Seafood Garden Restaurant	65.80
October 23	Avis Rent-a-Car	124.45
October 25	American Airlines	322.49
October 30	Dim Sum Chinese Restaurant	41.20

a. How much is the finance charge for October?

b. What is Eddie's new balance?

8. The Trust Bank of Salem offers a business line of credit which has an annual percentage rate of prime rate plus 3.6%, with a minimum of 10%. What is the APR if the prime rate is

a. 6.25% b. $9\frac{3}{4}$%

9. Crystal Enterprises has a $75,000 line of credit with Century National Bank. The annual percentage rate is the current prime rate plus $2\frac{1}{2}$%. The balance on April 1 was $32,200. On April 7, Crystal borrowed $10,800 to pay for a shipment of imported goods and on April 12 borrowed another $22,800 for warehouse renovations. On April 27, a $40,000 payment was made on the account. The billing cycle for April has 30 days. The current prime rate is $10\frac{1}{2}$%.

a. What is the finance charge on the account?

b. What is Crystal's new balance?

c. On May 1, how much credit does Crystal have left on the account?

10. The Happy Feet Shoe Stores chain accepts credit cards from customers who wish to charge their purchases. The fee that Visa charges Happy Feet is $3\frac{1}{4}\%$ of all transactions charged by customers using Visa cards. Last year Happy Feet did $275,500 in Visa business. How much was the fee for the year?

11. Roman Bas is interested in buying a van for a cash price of $9,899.00. He makes a 15% down payment and finances the balance with payments of $478.45 per month for 24 months.

a. What will be the amount of the finance charge on this loan?

b. What will be the total deferred payment price of the van?

12. Hawkeye Pierce, a professional photographer, found a camera he wanted to buy at Worldwide Foto. He had the option of paying $1,660 in cash or financing the camera with a $2\frac{1}{2}$ year installment loan. The loan required a 30% down payment and equal monthly payments of $73.21.

a. What is the finance charge on Hawkeye's loan?

b. What is the total deferred payment price of the camera?

13. Ivan Almas bought a riding lawnmower with a 12.3% add-on interest installment loan from The Friendly Finance Company. The purchase price of the mower was $3,270.00. Friendly required a 20% down payment and equal monthly payments for 48 months.

a. What is the total deferred payment price of the mower?

b. How much are Ivan's monthly payments?

14. Gary and Margie Robbins financed the construction of a swimming pool in their yard with a $7\frac{1}{2}$% add-on interest home improvement loan from Gary's credit union. The total price of the pool was $11,300.00. The credit union required equal monthly payments for 4 years.

a. What is the total deferred payment price of the pool?

b. How much are the monthly payments of the loan?

15. The Exotic Garden Flower shop purchased a delivery truck for $18,850. A down payment of $2,400 was made and the balance financed with monthly payments of $581.39 for 36 months.

a. What is the amount of the finance charge on the loan?

b. Use Table 13-1 to find what annual percentage rate was charged on the truck loan.

16. Santa Fe Industries purchased a $13,420 computer system from Computer City with a 48-month installment loan. The monthly payments are $412.88 per month.

a. Use the APR formula to calculate the annual percentage rate of the loan.

b. Use the APR tables to verify your answer from part a.

17. Autorama Car Sales uses the Beacon Bank to finance customer purchases. This month the bank is offering 36-month installment loans with an APR of 14.75%. For qualified buyers, a 10% down payment is required. Joan Sporn wants to buy a car for $15,490.

a. Use the APR tables to calculate the amount of the finance charge.

b. How much are the monthly payments on Joan's loan?

18. Find the sum of the digits of

a. 27 **b.** 36 **c.** 14 **d.** 60

19. **a.** What is the rebate fraction of a 24-month loan paid off after the 13th payment?

b. What is the rebate fraction of a 72-month loan paid off after the 39th payment?

20. The Campbell Corporation financed a new $35,000 conveyor belt for its production line with an $8\frac{1}{2}$% add-on interest installment loan for 48 months. The loan required a 10% down payment.

a. What is the amount of the finance charge on the loan?

b. How much are the monthly payments?

c. What annual percentage rate is being charged on the loan?

d. If Campbell decides to pay off the loan after 31 months, what is the loan payoff?

You are the newly hired assistant manager in the mortgage department of the Freedom Bank. As your first task your boss has asked you to calculate the required information for the following mortgages being considered by the bank:

	Amount Financed	Interest Rate	Term of Loan	Number of $1,000s Financed	Table Factor	Monthly Payment	Total Interest
1.	$99,000	8%	30 years	______	______	______	______
2.	$52,800	7%	25 years	______	______	______	______
3.	$170,700	$9\frac{1}{2}$%	15 years	______	______	______	______
4.	$229,600	$10\frac{3}{4}$%	30 years	______	______	______	______
5.	$1,000,000	$8\frac{1}{4}$%	15 years	______	______	______	______

6. Tamra Yoder purchased a home with a $122,000 mortgage at $8\frac{1}{2}$% for 25 years. Calculate the monthly payment and prepare an amortization schedule for the first 4 months of Tamra's loan.

Payment Number	Monthly Payment	Monthly Interest	Portion Used to Reduce Principal	Loan Balance
0				$122,000
1	______	______	______	______
2	______	______	______	______
3	______	______	______	______
4	______	______	______	______

7. Maria Ochoa bought a home with an adjustable rate mortgage. The margin on the loan is 4.5% and the rate cap is 7% over the life of the loan.

a. If the current index rate is 3.75%, what is the calculated interest rate of the ARM?

b. What is the maximum overall rate of Maria's loan?

8. For a $60,000, 10%, fixed-rate 30-year mortgage:

a. What is the amount of the first month's interest?

b. What portion of the first payment will be used to reduce the principal?

9. Sam Stubblefield purchased a summer home for \$78,200. He made a 25% down payment and financed the balance with a 15-year, $9\frac{1}{4}\%$ fixed-rate mortgage.

a. What is the amount of the monthly principal and interest portion of the loan?

b. Construct an amortization schedule for the first 5 months of Sam's mortgage:

Payment Number	Monthly Payment	Monthly Interest	Portion Used to Reduce Principal	Loan Balance
0				________
1	________	________	________	________
2	________	________	________	________
3	________	________	________	________
4	________	________	________	________
5	________	________	________	________

c. If the annual property taxes are \$1,370 and the hazard insurance premium is \$580 per year, what is the total monthly PITI of Sam's loan?

10. Suzanne Sitlington purchased a condominium for \$144,300. Her bank is willing to finance 80% of the purchase price. As part of the mortgage closing costs, Suzanne had to pay $2\frac{3}{4}$ discount points. How much does this amount to?

11. Dove Enterprises is looking for a 15-year mortgage to finance a \$3,000,000 warehouse addition. The Bentley Bank is offering a $9\frac{3}{4}\%$ mortgage with $3\frac{1}{4}$ discount points; Silver Savings and Loan is offering a 10% mortgage with 1 discount point. Dove is unsure which mortgage is a better deal and has asked for your help.

a. What is the total interest paid on each loan?

b. Taking into account the discount points, which lender is offering a better deal, and by how much?

12. Mitch Anderson is interested in a fixed-rate mortgage for \$56,000. He is undecided whether to choose a 20- or 25-year mortgage. The current mortgage rate is $10\frac{1}{2}\%$ for a 20-year mortgage and $11\frac{1}{2}\%$ for a 25-year mortgage.

a. What are the monthly principal and interest payments for each loan?

b. What is the total amount of interest paid on each loan?

c. Overall, how much more interest is paid by choosing the 25-year mortgage?

13. Butch Gemin owns a home that was recently appraised for $185,000. The balance on his existing first mortgage is $95,440. If Lake Tahoe Savings is willing to loan up to 75% of the appraised value, what is the potential amount of credit available to Butch on a home equity line of credit?

14. Davenport Truck Lines is planning to renovate its maintenance and repair facility. The current building was recently appraised at $550,000 and the balance on the existing first mortgage is $184,000. If Davenport's bank is willing to loan 80% of the appraised value, how much potential credit is available for the project?

15. Wilbur Chen is purchasing an $87,000 home. The down payment is 15% and the balance will be financed with a 30-year fixed-rate mortgage at $9\frac{1}{4}$% and $2\frac{3}{4}$ discount points. The annual property tax is $1,645 and the hazard insurance premium is $1,190. When Wilbur signed the original sales contract he put down a deposit of $6,000, which will be credited to his down payment. In addition, at the time of closing he must pay the following expenses:

Appraisal fee	$245.00
Credit report	$55.00
Attorney's fees	$310.00
Roof inspection	$70.00
Mortgage insurance premium	$\frac{3}{4}$% of amount financed
Termite inspection	$75.00
Title search	$138.00

As Wilbur's real estate broker, he has asked you the following:

a. What is the total monthly PITI of the mortgage loan?

b. What is the total amount of interest that will be paid on the loan?

c. How much is due from Wilbur at the time of the closing?

d. If the sellers are responsible for the broker's $5\frac{1}{2}\%$ commission, $1,150 in closing costs, and the existing mortgage with a balance of $25,000, what proceeds will be received on the sale of the property?

16. Donna Fletcher earns a gross income of $3,960.00 per month. She has submitted an application for a fixed-rate mortgage with a monthly PITI of $792.00. Donna has other financial obligations totaling $340.00 per month.

a. What is her housing expense ratio?

b. What is her total obligations ratio?

c. According to the lending ratio guidelines, for what type of mortgage would Donna qualify, if any?

Prepare the following statements.

1. **a.** Use the following financial information to calculate the owner's equity and prepare a balance sheet with vertical analysis as of December 31, 1996, for Party-Time Rentals, a sole proprietorship owned by Michael Brewley: Current assets, $232,540; property, plant, and equipment, $65,430; investments and other assets, $29,000; current liabilities, $69,600; long-term liabilities, $167,400.

Party-Time Rentals
Balance Sheet
December 31, 1996

b. The following financial information is for Party-Time Rentals as of December 31, 1997. Calculate the owner's equity for 1997 and prepare a comparative balance sheet with horizontal analysis for 1996 and 1997: Current assets, $265,300; property, plant, and equipment, $71,200; investments and other assets, $29,000; current liabilities, $51,500; long-term liabilities, $189,800.

Party-Time Rentals
Comparative Balance Sheet
December 31, 1996 and 1997

2. **a.** Use the following financial information to prepare a balance sheet with vertical analysis as of April 30, 1996, for Dollar Days Department Store, Inc.: Cash, $56,700; accounts receivable, $314,300; merchandise inventory, $248,500; prepaid insurance, $6,900; maintenance supplies, $2,330; land, $214,400; building, $717,000; fixtures, $101,900; delivery trucks, $73,900; computers, $43,700; goodwill, $80,000; investments, $122,400; accounts payable, $85,900; salaries payable, $38,200; notes payable (6-month), $152,000; mortgage payable, $494,800; debenture bonds, $138,000; common stock, $875,000; and retained earnings, $198,130.

Dollar Days Department Store, Inc.
Balance Sheet
April 30, 1996

b. The following financial information is for Dollar Days as of April 30, 1997. Prepare a comparative balance sheet with horizontal analysis for 1996 and 1997: Cash, $65,400; accounts receivable, $365,420; merchandise inventory, $213,700; prepaid insurance, $7,300; maintenance supplies, $3,420; land, $214,400; building, $815,000; fixtures, $132,760; delivery trucks, $68,700; computers, $42,100; goodwill, $90,000; investments, $145,600; accounts payable, $96,500; salaries payable, $33,600; notes payable (6-month), $143,120; mortgage payable, $481,300; debenture bonds, $138,000; common stock, $875,000; and retained earnings, $396,280.

Dollar Days Department Store, Inc.
Comparative Balance Sheet
April 30, 1996 and 1997

3. For the first quarter of 1996 Gemini Electronics had gross sales of $564,776, sales returns and allowances of $128,900, and sales discounts of $89,500. What were Gemini's net sales?

4. For the month of April, Southwest Auto Sales has the following financial information: Merchandise inventory, April 1, $1,443,510; merchandise inventory, April 30, $858,840; gross purchases, $487,210; purchase returns and allowances, $54,300; and freight in, $2,340.

a. What are Southwest's goods available for sale?

b. What is the cost of goods sold for April?

c. If net sales were $1,650,000, what was the gross margin for April?

d. If total operating expenses were $387,300, what was the net profit?

Prepare the following statements on separate sheets of paper.

5. **a.** From the following information for American Eagle Paper Company, construct an income statement with vertical analysis for the first quarter of 1997: Gross sales, $748,000; sales returns and allowances, $141,200; beginning inventory, Jan.1, $457,400; ending inventory, March 31, $517,300; net purchases, $279,600; total operating expenses, $328,000; income tax, $16,530.

American Eagle Paper Company
Income Statement
Jan. 1 – Mar. 31, 1997

b. From the second quarter 1997 data, prepare a comparative income statement with horizontal analysis for the first and second quarter of 1997: Gross sales, $784,330; sales returns and allowances, $167,400; beginning inventory, April 1, $425,430; ending inventory, June 30, $486,800; net purchases, $291,700; total operating expenses, $292,710; income tax, $36,200.

American Eagle Paper Company
Comparative Income Statement
First and Second Quarter, 1997

6. **a.** Use the following financial information to construct a 1996 income statement with vertical analysis for Shamrock Sporting Goods, Inc.: Gross sales, $3,010,000; sales returns and allowances, $254,700; sales discounts, $67,190; merchandise inventory, Jan. 1, 1996, $1,344,900; merchandise inventory, Dec. 31, 1996, $916,110; net purchases, $634,880; freight in, $21,320; salaries, $122,300; commissions, $379,900; rent, $285,100; depreciation, $77,300; utilities, $15,650; advertising, $441,690; insurance, $27,710; office expenses, $91,480; miscellaneous expenses, $88,300; and income tax, $24,290.

Shamrock Sporting Goods, Inc.
Income Statement
Jan. 1 – Dec. 31, 1996

b. The following data represents Shamrock's operating results for 1997. Prepare a comparative income statement with horizontal analysis for 1996 and 1997: Gross sales, \$3,129,800; sales returns and allowances, \$245,400; sales discounts, \$88,940; merchandise inventory, Jan. 1, 1997, \$1,698,800; merchandise inventory, Dec. 31, 1997, \$1,307,690; net purchases, \$723,230; freight in, \$31,210; salaries, \$103,600; commissions, \$411,120; rent, \$292,190; depreciation, \$60,050; utilities, \$17,530; advertising, \$440,900; insurance, \$25,200; office expenses, \$94,150; miscellaneous expenses, \$82,500; and income tax, \$33,590.

Shamrock Sporting Goods, Inc.
Comparative Income Statement
Jan. 1 – Dec. 31, 1996 and 1997

Calculate the amount of working capital and the current ratio for the following companies:

Company	Current Assets	Current Liabilities	Working Capital	Current Ratio
7. Quality Copiers, Inc.	\$546,700	\$322,180	______	______
8. Anchor Fence Company	121,650	75,400	______	______
9. Elite Beauty Supply	76,575	61,128	______	______

Use the additional financial information below to calculate the quick assets and acid test ratio for the companies in Questions 10–12.

Company	Cash	Marketable Securities	Accounts Receivable	Quick Assets	Acid Test Ratio
10. Quality Copiers, Inc.	\$54,390	\$72,300	\$181,200	______	______
11. Anchor Fence Company	15,400	0	74,230	______	______
12. Elite Beauty Supply	3,500	10,000	43,210	______	______

13. Calculate the average collection period for Quality Copiers, Inc., from Question 10 if the credit sales for the year amounted to \$1,384,000.

14. a. Calculate the average collection period for Elite Beauty Supply from Question 12 if the credit sales for the year amounted to \$243,700.

b. If last year's average collection period was 50 days, was the change good or bad?

c. If the industry average for similar-sized firms in the beauty supply business is 40 days, evaluate Elite's ratio.

Calculate the average inventory and inventory turnover for the following companies:

Company	Beginning Inventory	Ending Inventory	Average Inventory	Cost of Goods Sold	Inventory Turnover
15. Worldwide Electric	$237,600	$179,455	______	$1,642,000	______
16. Laurenzo's Market	452,338	613,080	______	6,440,200	______
17. Diversified Industries	10,290,127	12,059,570	______	21,398,620	______

18. The Five Star Awning Company had net sales of $2,655,000 last year. If the total assets of the company were $4,110,500, what was the asset turnover ratio?

Calculate the amount of owner's equity and the two leverage ratios for the following companies:

Company	Total Assets	Total Liabilities	Owner's Equity	Debt-to-Assets Ratio	Debt-to-Equity Ratio
19. Cellular Warehouse	$856,700	$546,680	______	______	______
20. Delta Investments	3,209,150	2,164,300	______	______	______
21. Budget Hardware, Inc.	131,200	39,760	______	______	______

Calculate the gross and net profits and the two profit margins for the following companies:

Company	Net Sales	Cost of Goods Sold	Gross Profit	Operating Expenses	Net Profit	Gross Profit Margin	Net Profit Margin
22. Skyline Roofing Co.	$915,200	$421,600	______	$373,460	______	______	______
23. Hydraulic Sales, Inc.	2,301,020	1,904,110	______	182,800	______	______	______
24. High-Tech Electronics	454,130	293,955	______	177,400	______	______	______

Using the owner's equity information below, calculate the return on investment for the companies in Questions 22–24:

	Owner's Equity	Return on Investment
25. Skyline Roofing Co.	$541,700	______
26. Hydraulic Sales, Inc.	822,900	______
27. High-Tech Electronics	125,340	______

28. Prepare a trend analysis chart from the following financial data for the Consolidated Equipment Company.

Consolidated Equipment Company
4-Year Selected Financial Data

	1996	1995	1994	1993
Net Sales	$1,767,800	$1,743,200	$975,900	$1,332,400
Net Income	132,120	121,200	16,200	98,550
Total Assets	3,220,700	3,114,500	3,031,000	2,910,040
Stockholder's Equity	536,000	521,600	419,650	501,230

Consolidated Equipment Company
Trend Analysis Chart

	1996	1995	1994	1993
Net Sales	______	______	______	______
Net Income	______	______	______	______
Total Assets	______	______	______	______
Stockholder's Equity	______	______	______	______

1. Calculate the total number of units available for sale and the cost of goods available for sale from the following inventory of 4′ × 8′ sheets of $\frac{1}{2}$″ plywood at the Liberty Lumber Yard:

Date	Units Purchased	Cost per Unit	Total Cost
Beginning Inventory, January 1	350	$6.50	______
Purchase, February 12	144	6.72	______
Purchase, June 5	175	6.20	______
Purchase, September 19	224	6.08	______
Purchase, November 4	311	6.47	______
Total units available	______	Cost of goods available for sale	______

2. When the manager of Liberty Lumber (Exercise 1) took physical inventory of the plywood on December 31, it was found that 260 sheets were on hand.

   a. What is the dollar value of the ending inventory using FIFO?

   b. What is the dollar value of the inventory using LIFO?

   c. What is the dollar value using the average cost method?

3. The following data represents the inventory figures for king-size mattresses at The Bedding Boutique.

The Bedding Boutique			Amount
January 1	Beginning inventory	230 units @ $167.05	______
May 19	Purchase	160 units @ $163.50	______
August 22	Purchase	135 units @ $166.80	______
December 12	Purchase	75 units @ $169.70	______
	Mattresses available for sale ______	Cost of goods available for sale	______

   a. How many mattresses did The Bedding Boutique have available for sale?

   b. What is the cost of the mattresses available for sale?

c. If physical inventory on December 31 showed 167 on hand, what is the value of those mattresses using FIFO?

d. What is the value of the 167 mattresses using LIFO?

e. What is the value of the mattresses using the average cost method?

4. Determine the value of the following inventory for Computer City using the lower-of-cost-or-market rule.

Description	Quantity	Unit Price Cost	Unit Price Market	Valuation Basis	Amount
16" Color Monitor	37	$545.60	$563.40	______	______
Model 24 Fax Modem	85	112.36	119.16	______	______
240 mb Hard Drive	63	232.28	213.52	______	______
350 mb Hard Drive	86	359.70	354.04	______	______
4 mb RAM Chip	230	128.40	122.00	______	______
8 mb RAM Chip	160	231.57	234.21	______	______
				Total value of inventory:	______

5. Using the retail method, estimate the value of the ending inventory at cost on June 30 from the following information for Winchester Distributors, Inc. Round the cost ratio to the nearest tenth percent.

Winchester Distributors, Inc.
June 1–June 30

	Cost	Retail
Beginning Inventory, June 1	$551,600	$788,000
Purchases	252,000	360,000
Net Sales $850,000		

6. Modern Office Products, Inc., maintains a gross margin of 63% on its executive line of desks and chairs. In April, Modern had a beginning inventory of $258,000, net purchases of $319,000, and net sales of $651,600. Use the gross profit method to estimate the cost of ending inventory.

7. The following data represents the inventory figures for Sheds America. Using the retail method, estimate the value of the ending inventory at cost on February 28. Round the cost ratio to the nearest tenth percent.

Sheds America
February 1–February 28

	Cost	Retail
Beginning Inventory, Feb. 1	$247,680	$550,400
Purchases (February)	193,950	431,000
Net Sales (February) $639,000		

8. On April 9, The Pacific Trading Company sustained a warehouse fire in which all of the merchandise was destroyed. Company records indicate that at the time of the fire the net sales to date were $1,315,800 and the purchases were $839,400. The beginning inventory, on January 1, was $518,100. For the past two years the company has maintained a gross margin of 44%. Use the gross margin method to estimate the inventory loss for the insurance claim.

9. The Sunglass Hut had net sales of $245,300 for the year. The beginning inventory at retail was $62,600 and the ending inventory at retail was $54,200.

   a. What is the average inventory at retail?

   b. What is the inventory turnover rounded to the nearest tenth?

10. Upwind Surfboards Company had cost of goods sold of $359,700 for the year. The beginning inventory at cost was $73,180 and the ending inventory at cost was $79,500.

   a. What is the average inventory at cost?

   b. What is the inventory turnover rate rounded to the nearest tenth?

11. Ventana Petroleum Corporation had cost of goods sold for the year of $3,890,000. The beginning inventory at cost was $375,600 and the ending inventory at cost amounted to $490,200. The inventory turnover rate published as the industry standard for a business of this size is 12 times.

a. Calculate the average inventory and actual inventory turnover rate for the company.

b. If the turnover rate is below 12 times, calculate the target average inventory needed to theoretically meet industry standards.

12. The Omega Pharmacy had net sales of $415,000 for the year. The beginning inventory at retail was $87,550, a mid-year inventory amounted to $94,500, and the ending inventory was $80,980. The inventory turnover rate published for a business of this size is 6.2 times.

a. Calculate the average inventory and actual inventory turnover rate for the company.

b. If the turnover rate is below 6.2 times, calculate the target average inventory to match the industry standard.

1. Arrow Industries purchased new production equipment totaling $437,000. Freight charges were $1,275 and installation amounted to $6,500. The equipment is expected to last 7 years and have a residual value of $13,000. If Arrow elects to use the straight-line method of depreciation, prepare a depreciation schedule for these assets.

Arrow Industries
Depreciation Schedule
Straight-line Method

End of Year	Annual Depreciation	Accumulated Depreciation	Book Value
		(new)	_______
1	_______	_______	_______
2	_______	_______	_______
3	_______	_______	_______
4	_______	_______	_______
5	_______	_______	_______
6	_______	_______	_______
7	_______	_______	_______

2. Capitol Paving, Inc., purchased a large asphalt truck for $85,000. It has an estimated useful life of 5 years and a trade-in value of $5,000. Using the sum-of-the-years' digits method, prepare a depreciation schedule for Capitol.

Capitol Paving, Inc.
Asphalt Truck, Depreciation Schedule

End of Year	Total Depreciation	Depreciation Rate Fraction	Annual Depreciation	Accumulated Depreciation	Book Value
				(new)	_______
1	_______	_______	_______	_______	_______
2	_______	_______	_______	_______	_______
3	_______	_______	_______	_______	_______
4	_______	_______	_______	_______	_______
5	_______	_______	_______	_______	_______

3. Delgado Construction Company bought air impact tools amounting to $23,500. They are expected to have a seven-year useful life and a trade-in value of $2,300. Prepare a depreciation schedule using the 200% declining-balance method for the tools.

Delgado Construction Company
Depreciation Schedule—Tools
7-year, 200% declining-balance

End of Year	Beginning Book Value	Depreciation Rate	Depreciation for the Year	Accumulated Depreciation	Ending Book Value
				(new)	_______
1	_______	_______	_______	_______	_______
2	_______	_______	_______	_______	_______
3	_______	_______	_______	_______	_______
4	_______	_______	_______	_______	_______
5	_______	_______	_______	_______	_______
6	_______	_______	_______	_______	_______
7	_______	_______	_______	_______	_______

4. Southeast Medical Center purchased a new x-ray machine for $625,000 with a salvage value of $20,000. Installation and calibration costs amounted to $3,500. The machine is expected to have a useful life of 20,000 hours of operation. Complete the following depreciation schedule using the units-of-production method:

Southeast Medical Center
Depreciation Schedule, Units-of-Production
X-ray Machine—20,000 Hours

End of Year	Depreciation per Hour	Hours	Annual Depreciation	Accumulated Depreciation	Book Value
				(new)	________
1	________	4,500	________	________	________
2	________	6,150	________	________	________
3	________	5,800	________	________	________
4	________	3,550	________	________	________

5. Bradco Supply Company purchased a new forklift for $26,500. The company expected the forklift to last six years or 30,000 miles, with an estimated salvage value of $1,200 at the end of that time. In the first year the forklift traveled 5,900 miles. As the accountant for Bradco, you are asked to determine which method of depreciation allows the *largest* write-off in the first year.

a. straight-line

b. sum-of-the-years' digits

c. double-declining balance

d. units-of-production

6. Dr. Anita Kim, a scientist at Georgia Tech, purchased research equipment for her laboratory amounting to $85,000. The setup and calibration expenses amounted to $6,500. Using Tables 17-1 and 17-2, what is the amount of the second year's depreciation under the modified accelerated cost recovery system?

7. General Motors purchased special tools amounting to $8,500,000. These tools are used in the assembly process of the Saturn automobile.

a. What is the property class for these tools under MACRS?

b. What is the percentage for the second year of depreciation for this property?

c. What is the amount of the depreciation expense in the final year of write-off?

8. Buttermilk Dairy Farms owns cattle valued at $970,000. As the accountant for the company, prepare a depreciation schedule for the cattle using MACRS.

Buttermilk Dairy Farms
Depreciation Schedule, MACRS
Dairy Cattle

End of Year	Original Basis (cost)	Cost Recovery Percentage	Cost Recovery (depreciation)	Accumulated Depreciation	Book Value

9. Southwest Investors Group purchased an oil well in Texas for $2,300,000. It is estimated that the well contains 1,400,000 barrels of oil. The residual value of the property and equipment is estimated to be $65,000.

a. What is the average depletion cost per barrel of oil?

b. If 135,000 barrels were pumped in the first year of operation, what is the amount of the depletion cost for that year?

1. Kelly Mitchell purchased a suit at Macy's for \$135.60. The alteration charge was \$18.50. The state sales tax is $5\frac{3}{4}\%$ and the county tax is 1.9%.

   a. What is the total amount of sales tax on the suit?

   b. What is the total purchase price?

2. Butch Porter bought auto parts for a total purchase price of \$227.90. The state has a 6% sales tax.

   a. What was the selling price of the auto parts?

   b. What was the amount of sales tax?

3. Last month Magic City Dry Cleaners had total sales, including sales tax, of \$47,837.50. The store is located in a state that has a sales tax of $7\frac{1}{2}\%$. As the accountant for Magic City, calculate:

   a. The amount of sales revenue for the shop last month.

   b. The amount of sales taxes that must be sent to the state Department of Revenue.

4. Recently Applied Technology Corporation purchased an executive jet for \$850,000. The state sales tax is 6.6% and the federal excise tax on airplanes is 10% on the amount over \$250,000.

   a. What is the amount of the sales tax?

b. What is the amount of the federal excise tax?

c. What is the total purchase price of the airplane?

5. The federal excise tax on ozone-depleting chemicals is $5.25 per pound for Halon and $1.75 per pound for CFC. If Igloo Air Conditioning purchased 650 pounds of Halon and 250 pounds of CFC last month, how much excise tax was due?

6. The Kramer Corporation purchased an office building with a market value of $3,154,800 in West Lake County. The assessment rate in the county is 100% and the tax rate is 12.37 mills.

a. What is the assessed value of the building?

b. What is the amount of property tax due?

7. Paul Davidson owns a house with an assessed value of $97,300. The tax rate is $24.30 per $1,000 of assessed value.

a. What is the amount of property tax due?

b. If the state offers a 3% discount for early payment, how much can Paul save by paying early?

c. If the state charges a mandatory $1\frac{1}{2}\%$ per month penalty for late payments, how much would the tax bill amount to if Paul paid 5 months late?

8. As the tax assessor for Summerville, you have been informed that an additional $750,000 in taxes will be required next year to build a new library and two parks. If the total assessed value of the property in Summerville is $850,000,000, how much will this add to current property taxes,

a. as a percent?

b. per $100 of assessed value?

c. per $1,000 of assessed value?

d. in mills?

Use the Tax Table, Exhibit 18-3, to calculate the tax liability for the following taxpayers earning under $100,000:

	Name	Filing Status	Taxable Income	Tax Liability
9.	Galahad	Married, Jointly	$32,119	_______
10.	Casey	Single	69,540	_______
11.	Lange	Married, Separately	74,210	_______
12.	Berns	Head of household	25,492	_______

Use the Tax Rate Schedules, Exhibit 18-4, to calculate the tax liability for the following taxpayers earning $100,000 or above:

	Name	Filing Status	Taxable Income	Tax Liability
13.	Levitt	Single	$205,439	_______
14.	Eastwood	Head of household	113,400	_______
15.	Antoine	Married, Separately	164,808	_______
16.	Wong	Married, Jointly	258,121	_______

As a newly hired IRS trainee, you have been asked to calculate the amount of tax refund or tax owed for the following taxpayers:

	Name	Tax Liability	Tax Credits	Other Taxes	Payments	(circle one) Refund/Owe	Amount
17.	Orlando	$4,100	$810	$593	$4,230	Refund/Owe	_______
18.	Chaplain	7,650	2,317	2,100	6,809	Refund/Owe	_______
19.	Major	7,439	0	0	7,540	Refund/Owe	_______
20.	Walker	16,187	3,429	1,820	13,690	Refund/Owe	_______

21. Austin Reed is a pilot for Delta Airlines. He is married, and for income tax purposes files jointly with his wife, Carrie. Together they claim five exemptions. Last year Austin earned $91,550 and Carrie earned an additional $12,700 at a part-time job. They lost $7,000 on the sale of an apartment building and had the following itemized deductions: $4,530 in real estate taxes, $6,210 in mortgage interest, $640 in charitable contributions, and $2,160 in unreimbursed job expenses above the 2% adjusted gross income exclusion. The Reeds are entitled to a tax credit of $2,135 and owe other taxes of $4,300. If they had a combined total of $16,200 withheld from their paychecks last year, calculate:

a. Adjusted gross income

b. Taxable income

c. Tax liability

d. Are the Reeds entitled to a refund or do they owe additional taxes? How much?

Calculate the amount of corporate income tax due and the net income after taxes for the following corporations:

	Name	Taxable Income	Tax Liability	Net Income after Taxes
22.	Aventura Corp.	$ 67,285	______	______
23.	Loehmans, Inc.	286,700	______	______
24.	Nordstrom, Inc.	17,900,000	______	______

1. Ross Kent is 25 years old and is interested in purchasing a 5-year term life insurance policy with a face value of $50,000. As his insurance agent, calculate the following insurance premiums for this policy.

   a. Annual

   b. Semiannual

2. Paul Burke is 40 years old and would like to purchase a 10-year term insurance policy with a face value of $5,000. Calculate his annual and monthly premiums.

   a. Calculate his annual premiums.

   b. Calculate his monthly premiums.

   c. How much more will Paul pay per year if he chooses monthly payments?

3. Donna Baum purchased a $125,000, 20-payment life insurance policy when she was 20 years old. She is now 40 years old and wishes to investigate her nonforfeiture options. As her insurance agent, calculate the value of Donna's three options.

   a. Option 1

   b. Option 2

c. Option 3

4. Orlando Lopez is evaluating his life insurance needs. His family's total living expenses are $31,100 per year. Carmen, his wife, earns $16,200 per year at her regular job and makes an additional $400 per month sewing and knitting. If the prevailing interest rate is $6\frac{1}{4}$%, how much life insurance should Orlando purchase to cover his dependents' income shortfall?

5. Tom Halloran, 40 years old, is considering purchasing either a whole life or a 20-payment life policy. If Tom is willing to spend $500 per year on insurance premiums, how much more coverage can he purchase under whole life than 20-payment life? Round to the nearest thousand.

6. You are the insurance agent for Kenwood International, a company that manufactures radio control equipment. The owner, Bill Jackson, would like you to give him a quote on the total premium for property insurance on a new production facility in the amount of $850,000, and equipment and contents valued at $4,600,000. The building is structural classification C and area rating 2.

7. National Distributors purchased property insurance to cover a large shipment of merchandise. The annual premium of the policy was $1,290. Four months later the goods were sold and National canceled the policy. What is the short-rate refund of premium?

8. Sunrise Building Supply Company has a property insurance policy with an annual premium of $1,750. In recent months, Sunrise failed 3 fire inspections due to missing or faulty fire extinguishers and a broken alarm system. After numerous warnings, the insurance company has elected to cancel the policy. If the policy has been in effect for 11 months, what is the regular refund due to Sunrise?

9. Allied Chemical had property valued at $250,000 and insured for $190,000. If the fire insurance policy contained an 80% coinsurance clause, how much would be paid by the insurance company in the event of a $65,000 fire?

10. Vanguard Auto Supplies, Inc., had multiple carrier fire insurance coverage in the amount of $840,000, as follows:

Equitable	$390,000	policy
Mutual of Omaha	$308,000	policy
Progressive	$142,000	policy
	$840,000	total coverage

Assuming that all coinsurance clause stipulations have been met, how much would each carrier be responsible for in the event of a $200,000 fire? (Round to the nearest whole percent.)

a. Equitable

b. Mutual of Omaha

c. Progressive

As an insurance agent, calculate the annual premium for the following clients:

	Name	Territory	Driver Class	Bodily Injury	Property Damage	Model Class	Vehicle Age	Comprehensive Deductible	Collision Deductible	Rating Factor	Annual Premium
11.	Lucas	4	2	100/300	50	D	0	$100	$250	2.0	______
12.	Molina	2	1	25/50	5	S	2	Full Coverage	250	1.4	______
13.	Bryant	3	3	10/20	10	A	4	Full Coverage	500	3.5	______
14.	Richards	1	4	15/30	100	M	6	$100	250	none	______
15.	Lee	2	4	50/100	25	Z	3	Full Coverage	500	4.0	______

16. Cheryl Hauser would like to purchase an automobile insurance policy with bodily injury and property damage coverage in the amounts of 50/100/25. In addition, she wants collision coverage with $500 deductible and comprehensive with $100 deductible. Cheryl is in driver classification 3, and lives in territory 1. Her vehicle, a Nissan Sentra, is in model class C and is 2 years old. Cheryl took driver training in high school and has not had a ticket or an accident in the past 5 years. Since she only drives her car for pleasure, the insurance company considers her to be a low risk and has assigned a rating factor of .85 to her policy. As her automobile insurance agent, calculate the total annual premium.

17. Vickie Stevenson's Pontiac was severely damaged by a mudslide in Malibu, California. The damage was estimated at $5,590. If Vickie carried full coverage comprehensive insurance and $250 deductible collision, what portion of the repair expenses will be paid by the insurance company?

18. Gordon Wardlow has automobile liability insurance in the amount of 15/30/10 and also carries $500 deductible collision coverage. Recently he was at fault in an accident in which his Chrysler collided with a Cadillac Eldorado. The three people in the Cadillac were injured. In the lawsuit settlement they were awarded the following: Steward, $20,000; Johnson $8,000; and Garfield, $11,000. Damages to the Cadillac amounted to $7,300 and damage to Gordon's car was $3,420.

a. How much will the insurance company have to pay, and to whom?

b. What part of the settlement will be Gordon's responsibility?

1. Bryant Corporation has 6,000,000 shares of common stock outstanding. If a dividend of $3,435,000 was declared by the company directors last year, what is the dividend per share of common stock?

2. The board of directors of National Dynamics has declared a dividend of $44,000,000. The company has 1,400,000 shares of preferred stock that pay $2.80 per share and 11,700,000 shares of common stock.

   a. What are the dividends due the preferred shareholders?

   b. What is the dividend per share of common stock?

3. Howell Chemical has 950,000 shares of $100 par value, $4\frac{1}{2}$%, cumulative preferred stock and 5,900,000 shares of common stock. Although no dividend was declared for the past two years, a $24,000,000 dividend has been declared for this year.

   a. How much is due the preferred shareholders?

   b. What is the dividend per share of common stock?

4. Southeast Foods is currently selling at $57\frac{3}{8}$. The earnings per share are $4.88 and the dividend is $1.30.

   a. What is the current yield of the stock?

   b. What is the price-earnings ratio?

5. You purchase 570 shares of Weldon Industries common stock at $\$54\frac{3}{4}$ per share. A few months later you sell the shares at $\$61\frac{7}{8}$. Your stockbroker charges 2% commission on round lots and an extra $1\frac{1}{2}\%$ on odd lots.

a. What is the total cost of the purchase?

b. What are the proceeds on the sale?

c. What is the gain or loss on the transaction?

6. On April 1, Joan Rhodes bought 15 Burdines bonds with a coupon rate of $8\frac{3}{8}$. The purchase price was $104\frac{7}{8}$ and the commission was \$5.00 per bond. Burdines bonds pay interest on March 1 and September 1.

a. What is the current yield of the bond?

b. What is the total purchase price of the bonds?

c. If Joan sold the bonds on December 1 for $99\frac{1}{8}$, what are the proceeds from the sale?

7. A mutual fund has an offer price of \$12.40 and a net asset value of \$11.95.

a. What is the sales charge?

b. What is the sales charge percent?

8. A mutual fund has total assets of $65,000,000 and liabilities of $4,500,000. If there are 9,400,000 shares outstanding, what is the net asset value of the fund?

9. Vince Edgar invested a lump sum of $30,000 in a mutual fund with an offer price of $13.90. How many shares did he purchase?

10. Bob Blumenkranz purchased 800 shares of Anaheim Fund for an offer price of $7.40 per share. He later sold the shares at a net asset value of $9.90 per share. During the time he owned the shares the fund paid dividends of $.43 and $.52 per share. What is Bob's return on investment?

1. You are the manager of the Main Street Playhouse. The following data represents the attendance figures for a five-day run of *The Phantom of the Opera,* a musical presented by a famous Broadway theater group.

Main Street Playhouse Attendance Figures

	Wednesday	Thursday	Friday	Saturday	Sunday
Matinee Performance	550	440	670	800	720
Evening Performance	600	520	780	990	910
Total Audience	1,150	960	1,450	1,790	1,630

a. Draw a single-line chart of the total audience for the week.

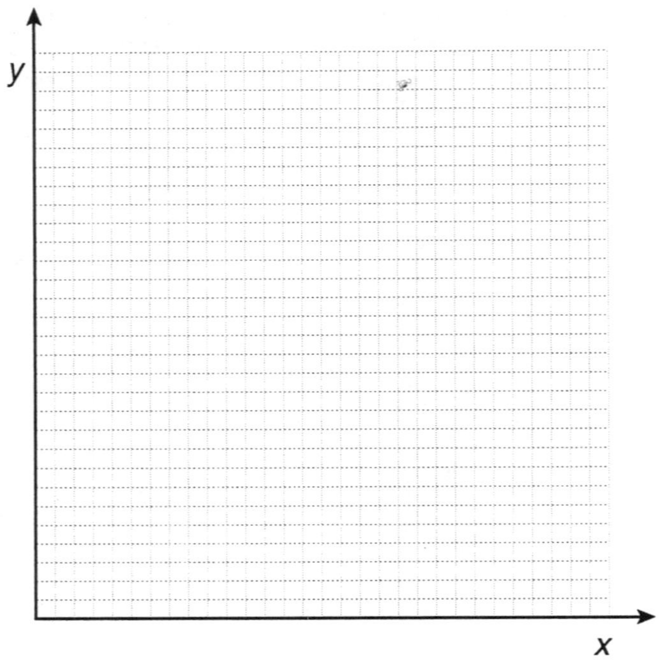

b. Draw a standard bar chart of the total audience for each day of the week.

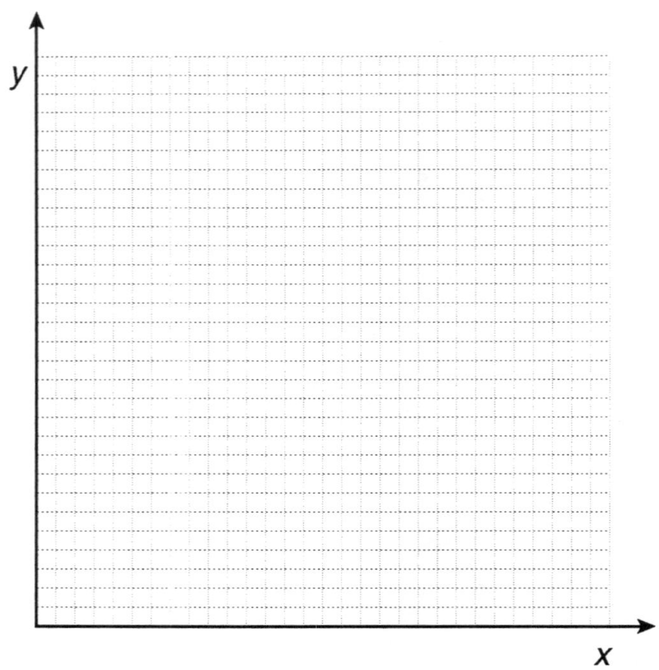

c. Draw a multiple-line chart showing the attendance at the matinee and evening performances. Use a solid line for the evening performance and a dashed line for the matinee.

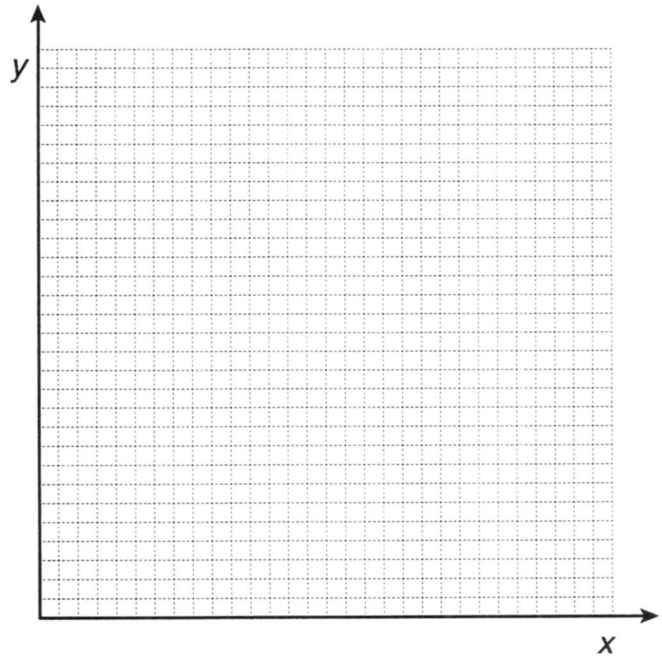

2. The following data reflects the sales and profit figures, in thousands, of The Dollar Days Department Store for June, July, and August. Dollar Days has branches in Woodside and Bayside. The merchandise is divided into three categories: hard goods, such as appliances and tools; soft goods, such as clothing and linens; and sporting goods.

Dollar Days Department Store: 3-Month Sales and Profits Report (in thousands)

		June		July		August	
		Woodside	**Bayside**	**Woodside**	**Bayside**	**Woodside**	**Bayside**
Hard goods:	**Sales**	56.5	74.3	61.0	77.1	48.8	75.5
	Profits	3.4	5.7	4.2	5.1	3.9	6.0
Soft goods:	**Sales**	62.3	75.1	58.8	71.9	50.0	79.5
	Profits	2.6	3.8	3.1	6.2	2.4	4.9
Sporting goods:	**Sales**	32.1	25.7	35.6	21.0	19.5	27.5
	Profits	1.1	1.8	1.2	1.3	.8	1.9

a. Construct a component bar chart displaying the hard goods sales of Woodside and Bayside as components of the total monthly sales of hard goods.

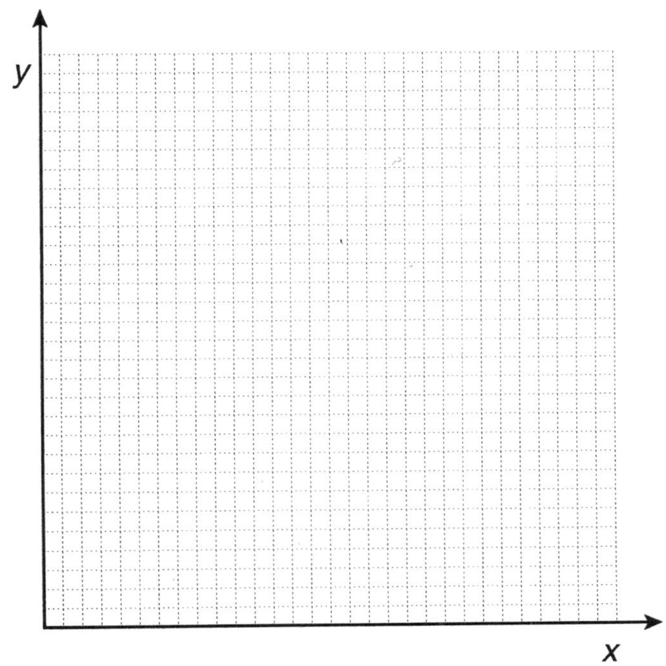

b. Construct a component bar chart displaying the sporting goods profits of Woodside and Bayside as components of the total monthly profits in sporting goods.

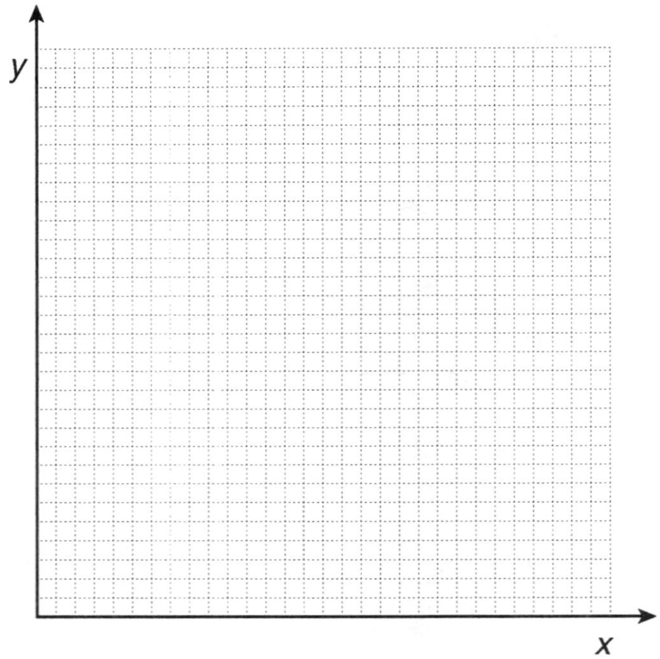

c. Construct a comparative bar chart for soft goods, comparing the monthly sales of the two stores.

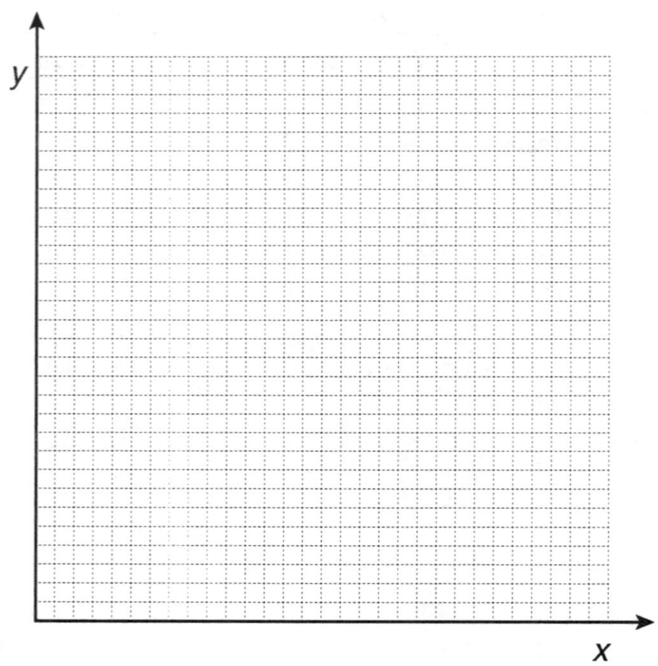

d. Construct a comparative bar chart comparing the total profits of each store for June, July, and August.

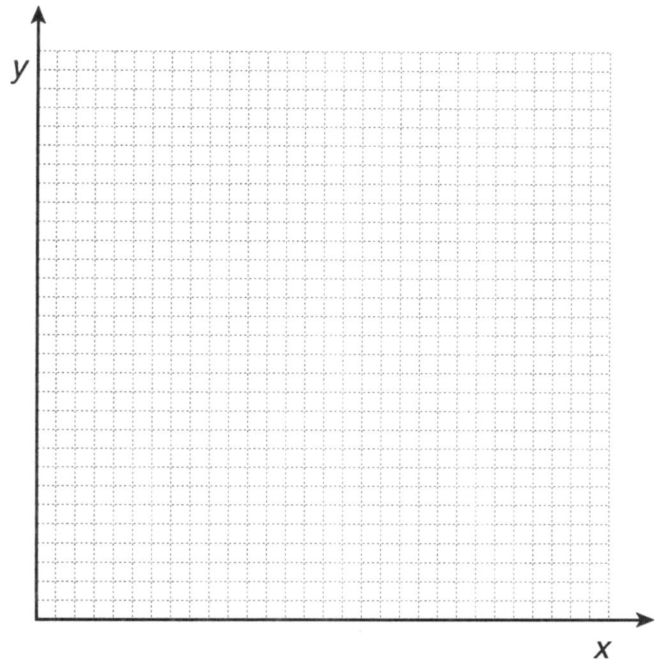

e. Construct a multiple-line chart displaying the total monthly sales for the three categories of merchandise. Make hard goods a solid line, soft goods a dashed line, and sporting goods a dotted line.

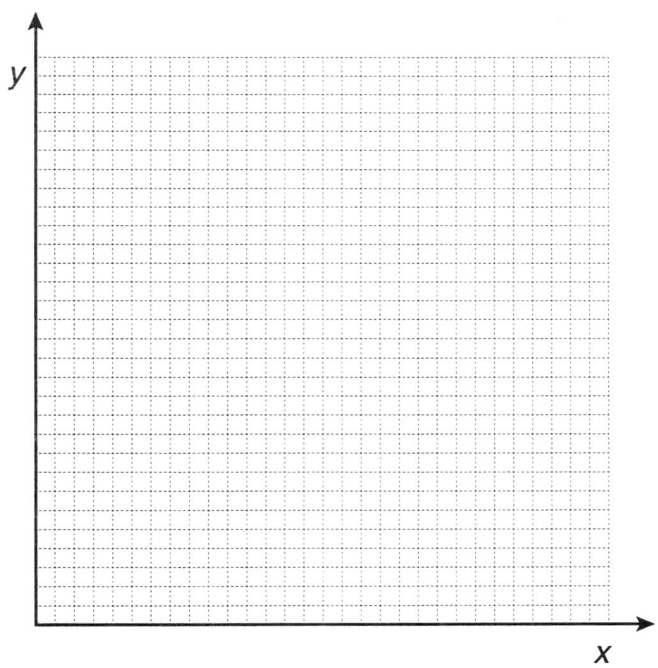

3. The population breakdown by age group of Bamster County is as follows:

Age	Percentage
Under 18	12.4
19–34	17.5
35–49	36.8
50–65	22.0
Over 65	11.3

As county manager, you have been asked by the mayor to construct a pie chart of this information to be presented at the next commission meeting.

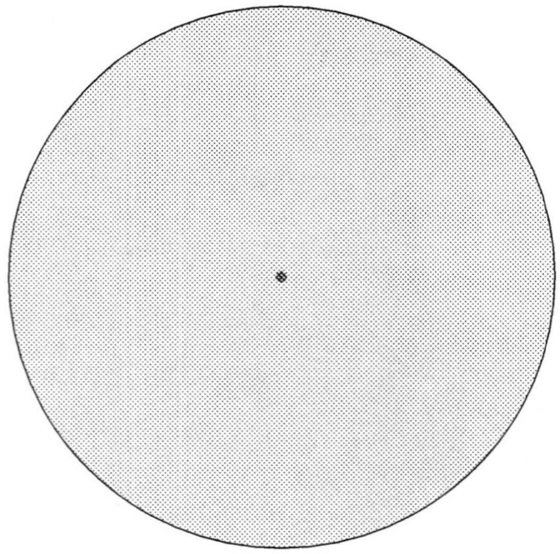

4. Stanford Manufacturing operates on three shifts per day. Yesterday shift 1 produced 167,000 units, shift 2 produced 125,000 units, and shift 3 produced 94,000 units. Convert each of these figures to percents of the total day's production, and construct a pie chart to show the results.

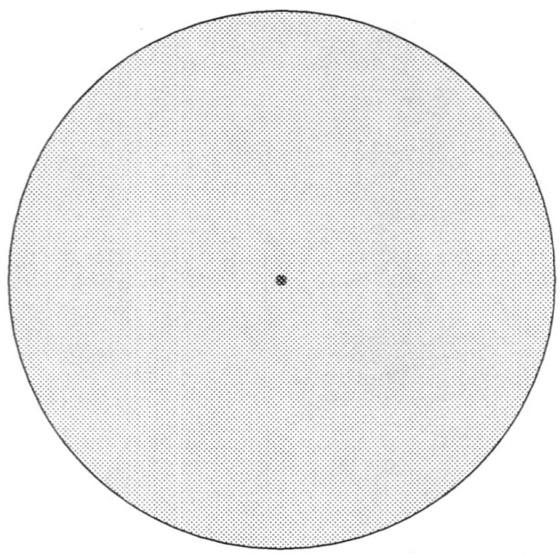

5. The following data represents the tons per day of corn delivered to a food processing plant for the month of June:

650	535	220	490	404	591	293	493	677	296
549	433	359	708	399	501	484	387	350	490
559	493	391	499	529	442	327	604	540	503

a. What is the mean of this set of data?

b. What is the median of this set of data?

c. What is the mode of this set of data?

d. What is the range of this set of data?

6. The Sweetie Pie Baking Company produced the following number of pies last week:

Mon: 335 Tues: 330 Wed: 293 Thurs: 251 Fri: 654 Sat: 267 Sun: 380

a. What is the mean of this set of data?

b. What is the median of this set of data?

c. What is the mode of this set of data?

d. What is the range of this set of data?

e. Which average best describes the production at Sweetie Pie? Why?

7. The following data represents the grades on a recent business math exam:

66 79 97 94 75 79 68 73 75 81 73 84 91 95 77 76 86 84 74 81 93
61 76 79 97 55 64 71 94 76 84 91 74 83 88 73 97 76 81 76 83 89

a. Group the grades into five classes of equal size (51–60, 61–70, 71–80, etc.) and construct a frequency distribution.

b. Calculate the mean of the grouped data using 55.5, 65.5, 75.5, etc. as the midpoints.

c. Prepare a histogram of the grade distribution.

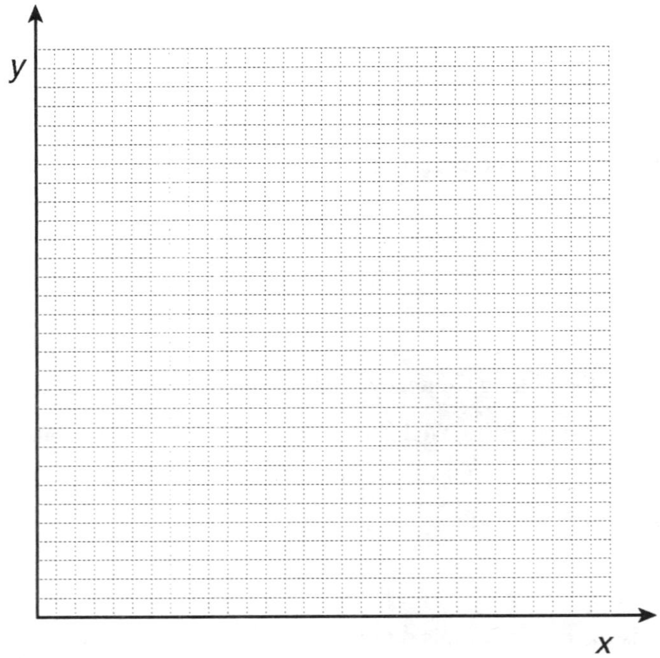

8. You are the manager of the All Star Roofing Company. The following set of data represents the number of square feet for each roof done by the company over the past month:

2,200	4,300	1,320	1,740	2,580	1,860	1,430	3,680	1,570	1,170	750	3,730
2,470	3,160	3,570	3,210	2,430	5,270	1,930	1,550	1,320	990	980	1,350

a. Group the data into six classes of equal size (0–999, 1,000–1,999, 2,000–2,999, etc.) and construct a frequency distribution of the roof data.

b. Calculate the mean of the grouped data using 499.5, 1,499.5, 2,499.5, etc. as the midpoints.

c. Prepare a histogram of this data to graphically illustrate your company's job sizes.

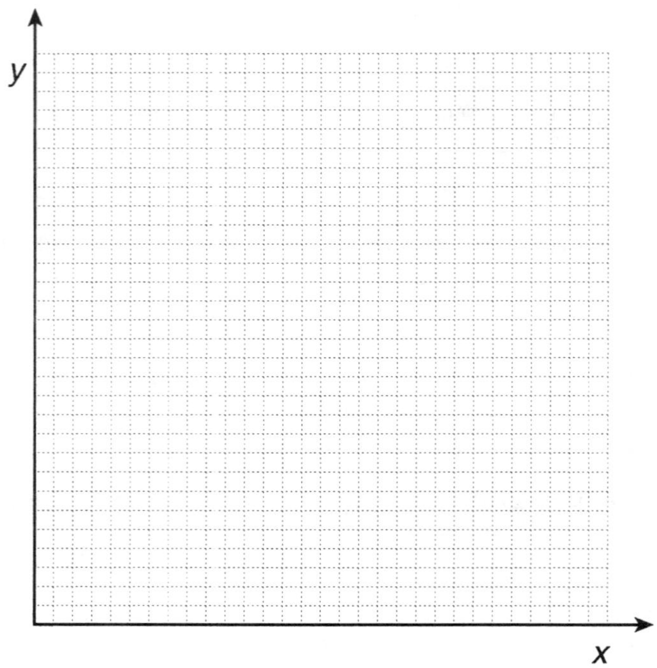

Business Measurements and Currency Conversion

Performance Objectives

SECTION I THE U.S. CUSTOMARY SYSTEM

1 Converting and solving problems involving length. (p. 133)
2 Converting and solving problems involving surface or square measure. (p. 133)
3 Converting and solving problems involving volume or capacity. (p. 134)
4 Converting and solving problems involving weight. (p. 136)
5 Converting and solving problems involving time. (p. 137)

SECTION II THE METRIC SYSTEM

6 Understanding and converting metric units of length, volume, and weight. (p. 138)
7 Understanding and converting metric units of surface area. (p. 140)
8 Converting and solving problems involving metric to U.S. and U.S. to metric measurements. (p. 141)
9 Converting and solving problems involving temperature. (p. 142)

SECTION III CURRENCY CONVERSION

10 Converting foreign currency to U.S. dollars and U.S. dollars to foreign currency (p. 144)

Measuring is the act of assigning a number to a quantity in order to represent its magnitude. It is one of the connecting links between mathematics and the physical world. Historically, units of measurement were related to various parts of the human body. The cubit, for example, was the length of the forearm from the elbow to the end of the middle finger, about 18 inches. The foot was said to be the length of a king's foot, varying over the years from 9 to 18 inches. Two steps of a soldier was a pace, and 1,000 paces was a mile. The weight of a pint of ale was called a pound, and thus the old saying, "A pint is a pound, the world around." As technology and trade progressed over time, more accurate and permanent standard units were established for weights and measures.

Historically, a mile measured 1,000 paces, or 2,000 steps by a marching soldier.

U.S. customary, or English system
A system of weights and measures, long used in the United States, using inches, feet, and yards for length; ounces, pounds, and tons for dry weight; and ounces, pints, quarts, and gallons for liquid measure.

metric system
The standard international metric system of measurement, known as SI, is a decimal system of weights and measures, based on 10. The basic units are the meter for length, the liter for volume, and the gram for weight.

Measurements are used in all aspects of business activity on a daily basis. Builders, architects, designers, and decorators are involved with measures of length and area. Warehouse managers and those involved in the shipping and handling of goods use weight and capacity calculations constantly. Managers of wholesale and retail businesses use surface measures such as floor space and shelf space allocations. In business, these calculations are often combined with dollar amounts to arrive at price per unit of measure, such as cost per square foot. Production lines use measurements of all kinds to manufacture an endless variety of products.

Until recently, the United States primarily recognized a cumbersome system of weights and measures known as the **U.S. customary** or **English system.** This system, brought to the United States by the colonists, uses the familiar units of yards, feet, and inches for length; ounces, pounds, and tons for dry weight; and ounces, pints, quarts, and gallons for liquid measure. In 1971, the United States began a gradual transition to a system used nearly worldwide known as the **metric system.** Although the U.S. government legalized the metric system for commerce and trade in 1866, it has not been widely accepted in this country.

Today, as more and more U.S. companies do business overseas, and an increasing number of foreign companies do business in the United States, it becomes necessary for all the countries of the world economic community to recognize and use the same system of measurements.

Aimed at helping U.S. industries stay competitive, the latest official U.S. government action regarding metrics is a provision within the American Technology Pre-eminence Act, signed into law in February 1992. The legislation includes amendments to the Fair Packaging & Labeling Act, requiring all packaging and labeling to use metric measurements for net quantities "as the primary system for measuring quantity." The use of ounces and pounds, commonly used as measurements in the past, has now become optional.

Section I deals with the U.S. customary, or English, system, which is still widely used in this country. Today's businessperson should still have a working knowledge of this system.

Section II covers the metric system of measurement, and the conversions between the U.S. and the metric systems. You need to be able to use the tables in this section to solve measurement problems and readily convert from one system to the other. Don't try to memorize these conversions, just remember where you put the tables!

Section III explains currency conversion; how to convert foreign currency to U.S. dollars and U.S. dollars to foreign currency. This important procedure is used by businesses and individuals involved in international investment, travel, and trade.

SECTION I

THE U.S. CUSTOMARY SYSTEM

The following measurement and conversion tables are for the U.S. customary system. They are separated into the categories of linear or straight measure; surface or square measure; volume, liquid, and dry measure; weight measure; and time.

The measurement tables are a complete listing for each category, while the conversion tables are for the most commonly used units. After looking them over, you will probably find that many of them are already familiar to you. Use the tables to solve the exercises at the end of each category. Your answers should be rounded to the nearest tenth.

1 Converting and Solving Problems Involving Length

Linear or **straight measure** is a one-dimensional unit, measuring the distance of something. Linear measurements are used in most businesses on a daily basis.

linear or straight measure
A one-dimensional unit, measuring the distance or length of something.

Measurement Table—Linear or Straight Measure

U.S. Customary Unit	U.S. Equivalents
inch (in.)	.083 foot
foot (ft)	$\frac{1}{3}$ yard, 12 inches
yard (yd)	3 feet, 36 inches
rod (rd)	$5\frac{1}{2}$ yards, $16\frac{1}{2}$ feet
furlong (fur)	$\frac{1}{8}$ mile, 220 yards
mile (mi) statute, land	1,760 yards, 5,280 feet
mile (nm) nautical, sea	1.151 statute miles

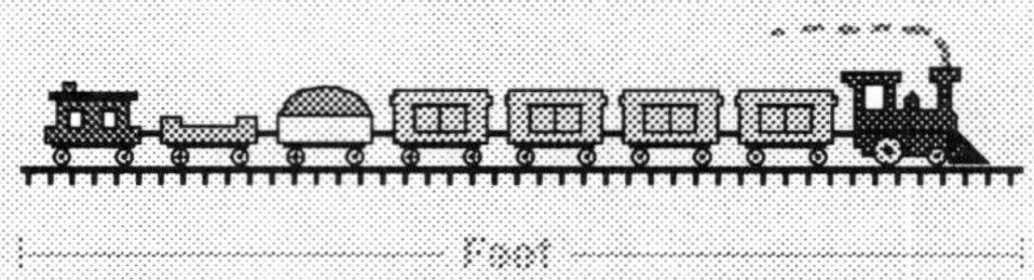

Conversion Table—Length

To Convert From	To	Multiply	To Convert From	To	Divide
feet	inches	number of feet by 12	inches	feet	number of inches by 12
yards	inches	number of yards by 36	inches	yards	number of inches by 36
yards	feet	number of yards by 3	feet	yards	number of feet by 3
miles	feet	number of miles by 5,280	feet	miles	number of feet by 5,280

EXAMPLE

How many feet are in 16 yards of electrical wire?

SOLUTION STRATEGY

To solve this problem, look up "yards to feet" in the conversion table. From the table we find this conversion is calculated by multiplying the number of yards by 3.

$$\text{Yards} \times 3 = \text{Feet} \quad 16 \times 3 = \underline{\underline{48 \text{ feet}}}$$

TRY-IT EXERCISES

Convert the following measurements as indicated:

1. 105 ft = __________ yd **2.** 156 in. = __________ ft

3. On a flight from Dallas to San Francisco, the pilot announces that the plane will be flying at 35,000 feet. How many miles high is the plane? (Round to a tenth of a mile.)

Check your answers with the solutions on page 147.

2 Converting and Solving Problems Involving Surface or Square Measure

Surface area or **square measure** represents the product of two dimensions, length and width. These measurements are labeled as square units, such as square inches or square feet. The formula for square measure is

surface area, or square measure
Measure representing the product of two dimensions, length and width, measured in square units.

$$\text{Area} = \text{Length} \times \text{Width}$$

Measurement Table—Surface or Square Measure

U.S. Customary Unit	U.S. Equivalents
square inch (sq in.)	.007 square feet
square foot (sq ft)	144 square inches
square yard (sq yd)	1,296 square inches 9 square feet
acre (a)	43,560 square feet 4,840 square yards
square mile (sq mi)	640 acres

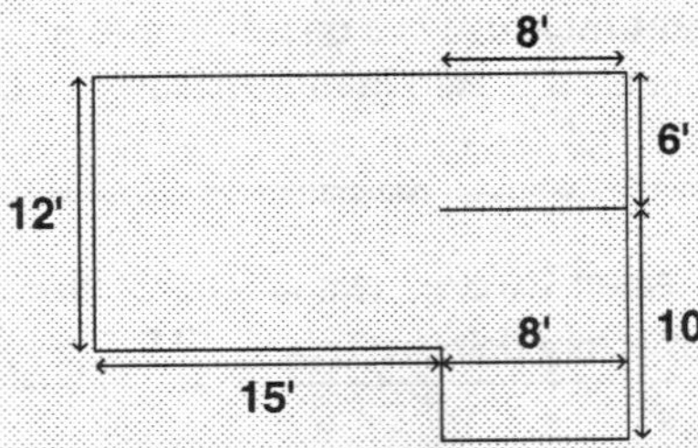

Conversion Table—Surface Area

To Convert From	To	Multiply
square feet	square inches	number of sq ft by 144
square yards	square feet	number of sq yd by 9
acres	square yards	number of acres by 4,840
square miles	acres	number of sq mi by 640

To Convert From	To	Divide
square inches	square feet	number of sq in. by 144
square feet	square yards	number of sq ft by 9
square yards	acres	number of sq yd by 4,840
acres	square miles	number of acres by 640

EXAMPLE

How many square yards are in 180 square feet of material?

SOLUTION STRATEGY

To solve this problem, look up "square feet to square yards" in the conversion table. From the table we find this conversion is calculated by dividing the number of square feet by 9.

$$\text{Square yards} = \frac{\text{Square feet}}{9} = \frac{180}{9} = \underline{\underline{20 \text{ square yards}}}$$

TRY-IT EXERCISES

Convert the following measurements as indicated:

4. 540 sq in. = __________ sq ft

5. 5 acres = __________ sq yd

6. How many square miles is a ranch containing 5,696 acres of land?

Check your answers with the solutions on page 147.

3 Converting and Solving Problems Involving Volume or Capacity

volume, or capacity
The size or extent of a three-dimensional object, measured in cubic units.

Volume or **capacity** is the size or extent of a three-dimensional object, measured in cubic units. It is calculated by the formula:

$$\text{Volume} = \text{Length} \times \text{Width} \times \text{Height}$$

Volume is divided into three categories: capacity, liquid measure, and dry measure. Capacity units are the measure of the size of a container, such as a cardboard box with 800 cubic inches of space, a railroad boxcar with 2,000 cubic feet of space, or a warehouse with 10,000 cubic yards of storage space. Liquid units are the measure of liquid quantities, such as quarts of milk, gallons of water, or barrels of oil. Dry units are the measure of non-liquid quantities, such as pints of strawberries, or bushels of wheat.

Measurement Table—Volume or Capacity Measure

U.S. Customary Unit	U.S. Equivalents
cubic inch (cu in.)	.00058 cubic foot
cubic foot (cu ft)	1,728 cubic inches
cubic yard (cu yd)	27 cubic feet

Liquid Measure

U.S. Customary Unit	U.S. Equivalents
fluid ounce (fl oz)	8 fluid drams
pint (pt)	16 fluid ounces
quart (qt)	2 pints
gallon (gal)	4 quarts
barrel (bbl)	varies from 31 to 42 gallons

Dry Measure

U.S. Customary Unit	U.S. Equivalents
pint (pt)	$\frac{1}{2}$ quart
quart (qt)	2 pints
peck (pk)	8 quarts
bushel (bu)	4 pecks

Conversion Table—Volume, Liquid, and Dry Measure

To Convert From	To	Multiply
Volume		
cubic feet	cubic inches	number of cu ft by 1,728
cubic yards	cubic feet	number of cu yd by 27
Liquid		
pints	fluid ounces	number of pints by 16
quarts	pints	number of quarts by 2
gallons	quarts	number of gallons by 4
Dry		
quarts	pints	number of quarts by 2
pecks	quarts	number of pecks by 8
bushels	pecks	number of bushels by 4

To Convert From	To	Divide
Volume		
cubic inches	cubic feet	number of cu in. by 1,728
cubic feet	cubic yards	number of cu ft by 27
Liquid		
fluid ounces	pints	number of fl oz by 16
pints	quarts	number of pints by 2
quarts	gallons	number of quarts by 4
Dry		
pints	quarts	number of pints by 2
quarts	pecks	number of quarts by 8
pecks	bushels	number of pecks by 4

EXAMPLE

How many quarts of milk can be packaged from a container tank holding 2,350 gallons?

SOLUTION STRATEGY

To solve this problem, look up "gallons to quarts" in the conversion table. From the table we find that this conversion is calculated by multiplying the number of gallons by 4.

Gallons × 4 = Quarts   2,350 × 4 = 9,400 quarts

TRY-IT EXERCISES

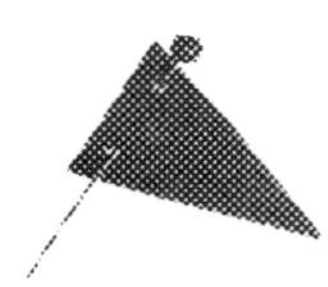

Convert the following measurements as indicated:

7. 80 fl oz = ________ pt   **8.** 12 pk = ________ qt

9. A large grocery store purchases a bushel of strawberries for \$45. The store sells them for \$1.19 per pint.

   a. How many pints of strawberries are in the bushel?

   b. How much gross profit is made if all the pints are sold?
   (Gross profit = Sales – Cost of merchandise sold)

Check your answers with the solutions on page 147.

4 Converting and Solving Problems Involving Weight

weight
The measure of the heaviness or mass of something.

Weight is the measure of the heaviness of something. These measurements are found extensively in the business world, particularly in the areas of production and distribution of goods. For example, a production manager may measure the output of a machine in pounds per hour, or the cost of steel in dollars per ton.

Measurement Table—Avoirdupois Weight

U.S. Customary Unit	U.S. Equivalents
grain (gr)	.036 dram
dram (dr)	27.344 grains
ounce (oz)	16 drams
pound (lb)	16 ounces
	7,000 grains
hundredweight (cwt)	100 pounds
ton, short (t)	2,000 pounds
ton, long	1.12 short tons
	2,240 pounds

Conversion Table—Weight

To Convert From	To	Multiply	To Convert From	To	Divide
pounds	ounces	number of pounds by 16	ounces	pounds	number of ounces by 16
tons (short)	pounds	number of tons by 2,000	pounds	tons (short)	number of pounds by 2,000
tons (long)	pounds	number of long tons by 2,240	pounds	tons (long)	number of pounds by 2,240

EXAMPLE

What is the weight in tons of 25,000 pounds of potatoes?

SOLUTION STRATEGY

To solve this problem, look up "pounds to tons" in the conversion table. From the table we find that this conversion is calculated by dividing the number of pounds by 2,000.

$$\frac{\text{Pounds}}{2{,}000} = \text{Tons (short)} \qquad \frac{25{,}000}{2{,}000} = \underline{\underline{12.5 \text{ tons}}}$$

TRY-IT EXERCISES

Convert the following measurements as indicated:

10. 48 oz = ________ lb **11.** 4 lb = ________ oz

12. If John weighs 173 lb 12 oz, and Sam weighs 88 oz less, how much does Sam weigh?

Check your answers with the solutions on page 147.

5 Converting and Solving Problems Involving Time

Time is the measure of the duration of an event. Time measurement and conversions are used continually, not only in business, but in people's everyday activities.

time
The measure of the duration of an event.

Measurement Table—Time

Standard Units	Equivalents
minute (min)	60 seconds
hour (hr)	60 minutes
day (da)	24 hours
week (wk)	7 days
month (mo)	$4\frac{1}{3}$ weeks
quarter (qtr)	13 weeks
	3 months
year (yr)	4 quarters
	12 months
	52 weeks
	365 days, 366 leap year
decade	10 years
century	10 decades
	100 years
millennium	100 decades
	1,000 years

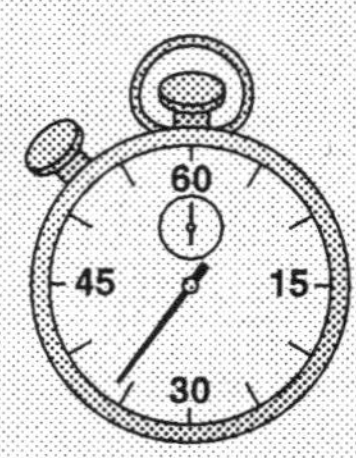

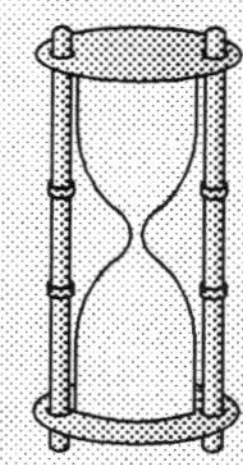

Conversion Table—Time

To Convert From	To	Multiply
minutes	seconds	number of minutes by 60
hours	minutes	number of hours by 60
days	hours	number of days by 24
weeks	days	number of weeks by 7
years	months	number of years by 12

To Convert From	To	Divide
seconds	minutes	number of seconds by 60
minutes	hours	number of minutes by 60
hours	days	number of hours by 24
days	weeks	number of days by 7
months	years	number of months by 12

EXAMPLE

If a project took 270 minutes to complete, how many hours did it take?

SOLUTION STRATEGY

To solve this problem, look up "minutes to hours" in the conversion table. From the table we find that this conversion is done by dividing the number of minutes by 60.

$$\frac{\text{Minutes}}{60} = \text{Hours} \qquad \frac{270}{60} = \underline{\underline{4.5 \text{ hours}}}$$

TRY-IT EXERCISES

Convert the following measurements as indicated:

13. 72 hr = __________ day **14.** 2.5 min = __________ sec

15. You have just received a loan from The Bank of America for 66 months. For how many years is this loan?

Check your answers with the solutions on page 147.

SECTION II

THE METRIC SYSTEM

The standard international metric system of measurement, known as SI, is a decimal system (based on the number 10) and is one of the easiest and most useful measuring systems ever established. A group of scientists developed the metric system after the French Revolution in 1789. Although it was slow to be accepted for many years, by 1875 it was adopted as the international standard of weights and measures.

6 Understanding and Converting Metric Units of Length, Volume, and Weight

meter (m)
The basic unit of measurement for length in the metric system.

liter (l)
The basic unit of measurement for volume in the metric system.

gram (g)
The basic unit of measurement for weight in the metric system.

The metric system is divided into three basic units: the meter, the liter, and the gram. The **meter (m)** is the basic unit of length and is slightly longer than a yard. The **liter (l)** is the basic unit of volume and is slightly larger than a quart. The **gram (g)** is the basic unit of weight and is about $\frac{1}{28}$ of an ounce. A penny weighs about $2\frac{1}{2}$ grams.

The metric system uses a series of prefixes to designate quantities larger or smaller than the basic units. Each quantity is a factor of 10 of the preceding prefix.

METRIC PREFIXES

For numbers **larger** than the basic unit, use the prefixes:

Prefix	Symbol	Meaning	Example
deca-	(dk)	10 times the basic unit	Example: 1 decameter (dkm) = 10 meters
hecto-	(h)	100 times the basic unit	Example: 1 hectoliter (hl) = 100 liters
kilo-	(k)	1,000 times the basic unit	Example: 1 kilogram (kg) = 1,000 grams

For numbers **smaller** than the basic unit, use the prefixes:

Prefix	Symbol	Meaning	Example
deci-	(d)	$\frac{1}{10}$ of the basic unit	Example: 1 deciliter (dl) = .1 of a liter
centi-	(c)	$\frac{1}{100}$ of the basic unit	Example: 1 centimeter (cm) = .01 of a meter
milli-	(m)	$\frac{1}{1,000}$ of the basic unit	Example: 1 milligram (mg) = .001 of a gram

Because the metric system is based on decimals, we convert to larger and smaller units by moving the decimal point to the right and to the left. Each place the decimal moves to the right is the same as multiplying by 10. Each place the decimal moves to the left is the same as dividing by 10.

Use the chart below and the following steps to convert between metric measurements.

STEPS FOR CONVERTING METRICS TO METRICS:

Step 1. Count the places between the two prefixes being converted.
Step 2. Move the decimal point that number of places to the left or right.

kilo-	hecto-	deca-	Basic Unit (meter, liter, gram)	deci-	centi-	milli-

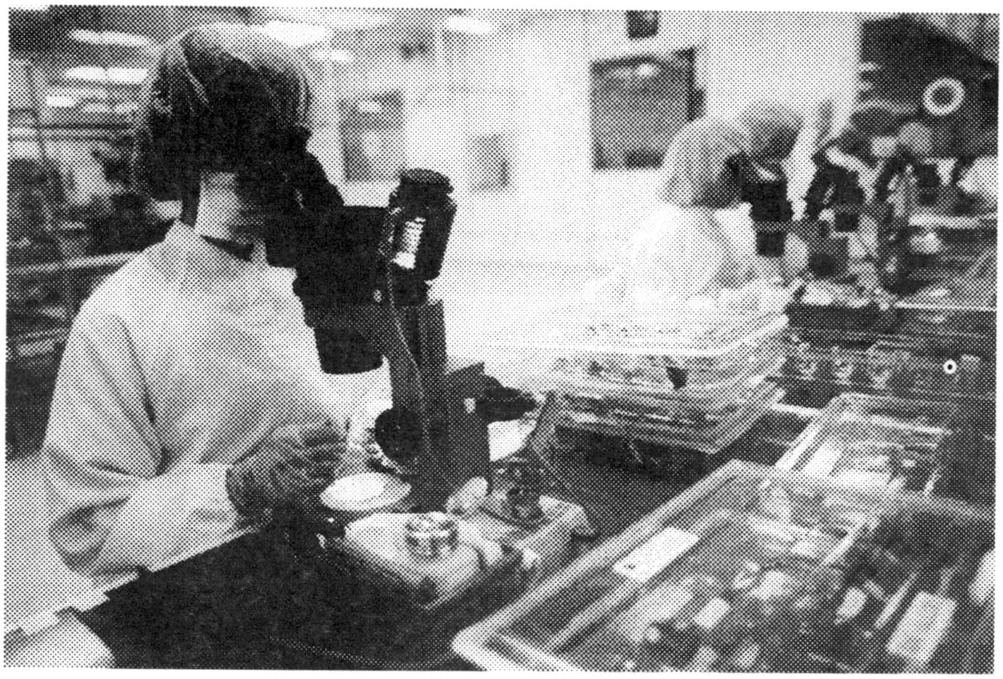

The metric system is the international standard of measurement used by scientists all over the world.

EXAMPLE

Convert 3.57 kilometers to meters.

SOLUTION STRATEGY

From the chart, a conversion from kilometers to the basic unit, meters, is a move of three places to the right:

1 2 3

kilo-	hecto-	deca-	Basic Unit Meters	deci-	centi-	milli-

Therefore, to convert a distance of 3.57 kilometers to meters, move the decimal point three places to the right.

3.57 kilometers = 3,570 meters

EXAMPLE

Convert 5,500 milligrams to grams.

SOLUTION STRATEGY

From the chart, a conversion from milligrams to grams is a move of three places to the left:

3 2 1

kilo-	hecto-	deca-	Basic Unit Grams	deci-	centi-	milli-

Therefore, to convert a weight of 5,500 mg to grams, move the decimal point three places to the left.

5,500 milligrams = 5.5 grams

TRY-IT EXERCISES

Convert the following as indicated:

16. 32 cm = ________ m

17. 12.5 kg = ________ g

18. 6.7 l = ________ ml

Check your answers with the solutions on page 147.

7 Understanding and Converting Metric Units of Surface Area

are (a)
A unit of measurement in the metric system used to measure surface area; it equals 100 square meters.

hectare (ha)
A unit of measurement in the metric system used to measure surface area; it is equal to approximately 2.5 acres.

In the metric system, the basic unit for surface measure is the square meter. Remember, when measuring surface area, we use the formula Area = Length × Width. One hundred square meters equal an **are (a).** An are is a surface measure of 10 meters by 10 meters. A unit known as a **hectare (ha)** is equal to 10,000 square meters, or a surface area of 100 meters by 100 meters. A hectare is approximately 2.5 acres.

Conversion Table—Metric Surface Area

To Convert From	To	Multiply
ares	square meters	number of ares by 100
hectares	ares	number of hectares by 100
hectares	square meters	number of hectares by 10,000
To Convert From	**To**	**Divide**
square meters	ares	number of square meters by 100
ares	hectares	number of ares by 100
square meters	hectares	number of square meters by 10,000

EXAMPLE

If a field measures 25,000 square meters, how many hectares is this?

SOLUTION STRATEGY

To solve this problem, look up "square meters to hectares" in the conversion table. From the table we find that this conversion is calculated by dividing the number of square meters by 10,000.

$$\text{Hectares} = \frac{\text{Square meters}}{10{,}000} = \frac{25{,}000}{10{,}000} = \underline{\underline{2.5 \text{ hectares}}}$$

TRY-IT EXERCISES

Convert the following as indicated:

19. 16.5 hectares = __________ square meters

20. 2,690 ares = __________ hectares

Check your answers with the solutions on page 147.

8 Converting and Solving Problems Involving Metric to U.S. and U.S. to Metric Measurements

The following alphabetical conversion table can be used to convert the most common U.S. and metric units. Note that the math procedure for each conversion is multiplication. Remember, don't try to memorize the conversion factors; learn to use the conversion tables as your tools! In everyday business, the more frequently you use various measurements, both U.S. and metric, the more familiar they will become.

Conversion Table—Metric to U.S. and U.S. to Metric

To Convert From	To	Multiply	To Convert From	To	Multiply
acres	hectares	acres by .4047	meters	feet	meters by 3.2808
ares	square yards	ares by 119.6	meters	inches	meters by 39.370
bushels	liters	bushels by 35.238	meters	yards	meters by 1.094
centimeters	inches	centimeters by .3937	miles	kilometers	miles by 1.6093
centimeters	yards	centimeters by .01094	ounces (dry)	grams	ounces by 28.35
cubic centimeters	cubic inches	cubic centimeters by .061	ounces (fluid)	liters	ounces by .0296
cubic inches	cubic centimeters	cubic inches by 16.3872	pecks	liters	pecks by 8.8096
cubic inches	liters	cubic inches by .0164	pints (dry)	cubic centimeters	pints by 550.41
cubic yards	cubic meters	cubic yards by .7656	pints (fluid)	cubic centimeters	pints by 473.179
feet	meters	feet by .3048	pounds	kilograms	pounds by .4536
gallons	liters	gallons by 3.7854	quarts (dry)	cubic centimeters	quarts by 1101.23
grams	ounces	grams by .0353	quarts (fluid)	liters	quarts by .9463
hectares	acres	hectares by 2.471	square centimeters	square inches	square centimeters by .155
inches	centimeters	inches by 2.54	square feet	square meters	square feet by .0929
kilograms	pounds	kilograms by 2.2046	square inches	square centimeters	square inches by 6.4517
kilograms	tons	kilograms by .0011	square meters	square inches	square meters by 1550.0
kilometers	miles	kilometers by .6214	square meters	square feet	square meters by 10.764
liters	cubic inches	liters by 61.025	square meters	square yards	square meters by 1.196
liters	gallons	liters by .2642	square yards	square meters	square yards by .836
liters	quarts (dry)	liters by .9081	tons (short)	kilograms	tons by 907.1846
liters	ounces (fluid)	liters by 33.7838	tons (short)	tons (metric)	tons (short) by .9072
liters	quarts (fluid)	liters by 1.0567	tons (metric)	tons (short)	tons (metric) by 1.1023
			yards	meters	yards by .9144

EXAMPLE

How many liters of gasoline will be required to fill a 12-gallon tank? Round to the nearest tenth.

SOLUTION STRATEGY

To solve this problem, look up "gallons to liters" in the conversion table. From the table we find that this conversion is calculated by multiplying the number of gallons by 3.7854.

Liters = Gallons × 3.7854 Liters = 12 × 3.7854 = 45.4 liters

TRY-IT EXERCISES

Convert the following, as indicated (round answers to the nearest tenth where needed):

21. 69 pounds = _______ kilograms

22. 800 square meters = _______ square feet

23. 15 meters = _______ inches

24. 30 liters = _______ quarts (dry)

Check your answers with the solutions on page 147.

9 Converting and Solving Problems Involving Temperature

temperature
A measure of the warmth or coldness of an object or substance.

Fahrenheit (F)
The basic unit of measurement for temperature in the U.S. customary system.

Celsius (C)
The basic unit of measurement for temperature in the metric system.

Temperature is a measure of the warmth or coldness of an object or substance. The U.S. customary system measures temperature in degrees **Fahrenheit (F),** while the metric system uses degrees **Celsius (C).** In the Celsius temperature scale, 0 degrees is the freezing point of water, and 100 degrees is the boiling point of water.

To convert from Celsius to Fahrenheit, use the formula:

$$F = 1.8C + 32$$

To convert from Fahrenheit to Celsius, use the formula:

$$C = \frac{5}{9}(F - 32)$$

EXAMPLE

How many degrees Fahrenheit is 38.5° Celsius?

SOLUTION STRATEGY

To convert from Celsius to Fahrenheit use the formula:

$$F = 1.8C + 32$$
$$F = 1.8(38.5) + 32$$
$$F = 69.3 + 32$$
$$F = \underline{\underline{101.3^\circ}}$$

EXAMPLE

How many degrees Celsius is 302° Fahrenheit?

SOLUTION STRATEGY

To convert from Fahrenheit to Celsius use the formula:

$$C = \frac{5}{9}(F - 32)$$
$$C = \frac{5}{9}(302 - 32)$$
$$C = \frac{5}{9}(270)$$
$$C = \underline{\underline{150^\circ}}$$

TRY-IT EXERCISES

25. 50°C = _________ °F   **26.** 140°F = _________ °C

Check your answers with the solutions on page 147.

SECTION III
CURRENCY CONVERSION

United States dollars are legal **currency,** or **money,** only in the United States. International investment, travel, and trade require that dollars be exchanged for foreign currency. Price quotations of international transactions are generally expressed in the seller company's national currency. U.S. exports, for example, are quoted in U.S. dollars ($); British exports are quoted in pounds (£); German exports in marks (DM), and so on. A Japanese company purchasing goods from an American company would have to exchange Japanese yen (¥) for U.S. dollars before purchasing the goods.

currency, or money
A country's legally established medium of exchange and a standard of value for its goods, services, and assets.

Price quotations are generally expressed in the vendor company's national currency.

exchange rate
A currency's value expressed in terms of another country's currency.

floating exchange rate
Today's worldwide exchange rate system in which rates are allowed to vary on a daily basis among major currencies.

transaction exposure
In international trade, the time period between the signing of the contract and the delivery of the goods, in which importers and exporters are *exposed* to the rise and fall of currency exchange rates.

Most countries have their own monetary system and currency. Each country's money acts as a medium of exchange and a standard of value for its goods, services, and assets. The **exchange rate** of a currency is its legally established value expressed in terms of another country's currency. For example, if it takes 105 Japanese yen to purchase 1 U.S. dollar, this would be written as: ¥105 to $1 U.S.

Until 1973, exchange rates between currencies from around the world were fixed at a specified amount. Fixed rates made international currency exchanges quite routine, but did not reflect the economic and political events and pressures occurring daily around the world. Today we have a system known as **floating exchange rates** in which rates are allowed to vary on a daily basis among major currencies.

Floating exchange rates are a particular threat to companies entering into international agreements that may take two or three months or even longer to complete. From the time the contract is signed to the time the goods are delivered there is **transaction exposure.** When the home currency is weak against the other currency of the transaction, the company whose currency is being exchanged is at risk of incurring exchange rate losses. On the other hand, when the home currency is strong against the other currency, extra gains could be realized. As an inducement to closing a sale, sellers sometimes agree to do business in the buyer's national currency, thus shifting the exposure to rate changes from the buyer to the seller.

Let's look at an example of how exchange rate fluctuations affect a transaction. A U.S. company places an order for machine parts from a firm in Japan. It is agreed that the transaction price will be 10,000,000 Japanese yen. At the time the order was written the exchange rate was .0095 dollars per yen. The U.S. import manager calculated that the order would cost $95,000.

$$10{,}000{,}000 \text{ yen} \times .0095 \text{ dollars per yen} = \underline{\$95{,}000}$$

When the shipment arrived from Japan 60 days later and the payment was due, the exchange rate had changed to $.0098 per yen. It now took more of a dollar to purchase a yen. The dollar is said to have weakened against the yen. The U.S. company would now have to spend $98,000 to purchase the 10,000,000 yen needed to pay the bill.

$$10{,}000{,}000 \text{ yen} \times .0098 \text{ dollars per yen} = \underline{\$98{,}000}$$

If on the other hand, the value of the dollar had strengthened to say $.0092, then less of a dollar is necessary to purchase each yen. The U.S. company would have to pay only $92,000 for the order.

$$10{,}000{,}000 \text{ yen} \times .0092 \text{ dollars per yen} = \underline{\$92{,}000}$$

Successfully managing currency risk is very important to companies engaged in international trade. With today's global marketplace and intense competition, a company cannot simply pass the added cost of foreign exchange losses on to the customer as higher prices.

When the delivery time of an order is over 30 days, a **forward contract** can be used to reduce the risk of losses. A forward contract is an agreement between the exporter and the importer to exchange currencies on a specified date in the future, at a price agreed upon today. These contracts can be written for 30, 90, or 180 days into the future. The cost of a forward contract is the difference between the cost of the **spot market,** today's cost of money, and the cost of the **forward market.** These contracts are arranged by international banks and brokerage houses around the world.

forward contract
A means of reducing the risk of transaction exposure, whereby the importer and exporter agree to exchange currencies on a specified date in the future, at a price agreed upon today.

spot market
Today's currency exchange rate, or cost of money.

forward market
The currency exchange rate, or cost of money, on a specified date in the future, usually 30, 90, or 180 days.

10 Converting Foreign Currency to U.S. Dollars and U.S. Dollars to Foreign Currency

Billions of dollars, in over 50 foreign currencies, are exchanged in world trade and investment each day. Currency exchange rate tables are published daily in the business section of most major newspapers. As an example, let's take a column-by-column look at a typical day's listing for Germany. Exhibit 1 is such a table, reprinted from *The Wall Street Journal.*

Column 1	*Germany (Mark)* This column lists the countries in alphabetical order followed by the name of each currency, in parentheses. For Germany, 30-, 90-, and 180-day forward contracts are also listed.
Columns 2 & 3	*U.S. $ equiv.* These columns list the exchange rates for the past two days expressed as "U.S. dollars per unit of foreign currency."
Column 2	*Tuesday*—It took .5976 U.S. dollars to purchase one mark.
Column 3	*Monday*—It took .6051 U.S. dollars to purchase one mark. Since it required *less* of a dollar on Tuesday than on Monday to purchase one mark, the U.S. dollar is said to have *strengthened* against the German mark.
Columns 4 & 5	*Currency per U.S. $* These columns list the exchange rates for the past two days expressed as "units of foreign currency per U.S. dollar."
Column 4	*Tuesday*—It took 1.6735 German marks to purchase one U.S. dollar.
Column 5	*Monday*—It took 1.6525 German marks to purchase one U.S. dollar. Since it required more marks on Tuesday than on Monday to purchase one dollar, the German mark is said to have *weakened* against the U.S. dollar.

CURRENCY TRADING

EXCHANGE RATES

Tuesday, May 10, 1994

The New York foreign exchange selling rates below apply to trading among banks in amounts of $1 million and more, as quoted at 3 p.m. Eastern time by Bankers Trust Co., Dow Jones Telerate Inc. and other sources. Retail transactions provide fewer units of foreign currency per dollar.

1	2	3	4	5
	U.S. $ equiv.		Currency per U.S. $	
Country	Tues.	Mon.	Tues.	Mon.
Argentina (Peso)	1.01	1.01	.99	.99
Australia (Dollar)	.7237	.7207	1.3818	1.3875
Austria (Schilling)	.08496	.08606	11.77	11.62
Bahrain (Dinar)	2.6522	2.6522	.3771	.3771
Belgium (Franc)	.02903	.02940	34.45	34.02
Brazil (Cruzeiro real)	.0006852	.0006966	1459.36	1435.48
Britain (Pound)	1.4885	1.4980	.6718	.6676
30-Day Forward	1.4875	1.4970	.6723	.6680
90-Day Forward	1.4871	1.4964	.6724	.6683
180-Day Forward	1.4868	1.4963	.6726	.6683
Canada (Dollar)	.7262	.7237	1.3770	1.3817
30-Day Forward	.7252	.7227	1.3790	1.3837
90-Day Forward	.7232	.7206	1.3827	1.3878
180-Day Forward	.7204	.7174	1.3882	1.3940
Czech. Rep. (Koruna)				
Commercial rate	.0341111	.0341111	29.3160	29.3160
Chile (Peso)	.002400	.002400	416.73	416.73
China (Renminbi)	.114943	.114943	8.7000	8.7000
Colombia (Peso)	.001189	.001189	841.35	841.35
Denmark (Krone)	.1527	.1546	6.5476	6.4679
Ecuador (Sucre)				
Floating rate	.000466	.000466	2146.01	2146.01
Finland (Markka)	.18414	.18617	5.4305	5.3715
France (Franc)	.17438	.17648	5.7345	5.6665
30-Day Forward	.17418	.17626	5.7411	5.6735
90-Day Forward	.17402	.17610	5.7466	5.6785
180-Day Forward	.17403	.17611	5.7463	5.6783
Germany (Mark)	.5976	.6051	1.6735	1.6525
30-Day Forward	.5971	.6046	1.6749	1.6539
90-Day Forward	.5969	.6045	1.6752	1.6542
180-Day Forward	.5978	.6056	1.6727	1.6513
Greece (Drachma)	.004056	.004104	246.55	243.65
Hong Kong (Dollar)	.12943	.12942	7.7260	7.7265
Hungary (Forint)	.0097828	.0097991	102.2200	102.0500
India (Rupee)	.03212	.03212	31.13	31.13
Indonesia (Rupiah)	.0004630	.0004630	2160.01	2160.01
Ireland (Punt)	1.4520	1.4641	.6887	.6830
Israel (Shekel)	.3328	.3328	3.0050	3.0050
Italy (Lira)	.0006257	.0006304	1598.21	1586.42
Japan (Yen)	.009574	.009742	104.45	102.65
30-Day Forward	.009593	.009762	104.24	102.44
90-Day Forward	.009635	.009806	103.79	101.98
180-Day Forward	.009709	.009884	103.00	101.17
Jordan (Dinar)	1.4620	1.4620	.6840	.6840
Kuwait (Dinar)	3.3599	3.3599	.2976	.2976
Lebanon (Pound)	.000592	.000592	1689.00	1689.00
Malaysia (Ringgit)	.3820	.3831	2.6180	2.6105
Malta (Lira)	2.5954	2.5954	.3853	.3853
Mexico (Peso)				
Floating rate	.2994012	.2994012	3.3400	3.3400
Netherland (Guilder)	.5322	.5390	1.8788	1.8554
New Zealand (Dollar)	.5815	.5790	1.7197	1.7271
Norway (Krone)	.1378	.1395	7.2555	7.1710
Pakistan (Rupee)	.0331	.0331	30.24	30.24
Peru (New Sol)	.4726	.4726	2.12	2.12
Philippines (Peso)	.03761	.03761	26.59	26.59
Poland (Zloty)	.00004466	.00004452	22389.00	22460.00
Portugal (Escudo)	.005787	.005859	172.79	170.67
Saudi Arabia (Riyal)	.26665	.26665	3.7502	3.7502
Singapore (Dollar)	.6421	.6456	1.5575	1.5490
Slovak Rep. (Koruna)	.0308261	.0307314	32.4400	32.5400
South Africa (Rand)				
Commercial rate	z	.2775	z	3.6040
Financial rate	z	.2116	z	4.7250
South Korea (Won)	.0012372	.0012407	808.30	806.00
Spain (Peseta)	.007274	.007359	137.48	135.89
Sweden (Krona)	.1290	.1301	7.7491	7.6850
Switzerland (Franc)	.6981	.7090	1.4325	1.4105
30-Day Forward	.6983	.7091	1.4320	1.4102
90-Day Forward	.6996	.7104	1.4293	1.4076
180-Day Forward	.7024	.7134	1.4237	1.4017
Taiwan (Dollar)	.037526	.037526	26.65	26.65
Thailand (Baht)	.03968	.03968	25.20	25.20
Turkey (Lira)	.0000286	.0000287	34991.35	34862.11
United Arab (Dirham)	.2730	.2730	3.6628	3.6628
Uruguay (New Peso)				
Financial	.200602	.200602	4.99	4.99
Venezuela (Bolivar)				
Floating rate	.00757	.00757	132.13	132.13
SDR	1.41326	1.41754	.70758	.70545
ECU	1.15230	1.16490	....	....

Special Drawing Rights (SDR) are based on exchange rates for the U.S., German, British, French and Japanese currencies. Source: International Monetary Fund.

European Currency Unit (ECU) is based on a basket of community currencies.

z-Not quoted.

STEPS TO CONVERT CURRENCY USING THE EXCHANGE RATES TABLE:

A. Foreign Currency to U.S. Dollars:

Step 1. Locate the exchange rate listed in Column 2, *U.S. $ equiv.*, for the country in question.

Step 2. Multiply the number of units of foreign currency by the exchange rate for that currency:

U.S. dollars = Units of foreign currency × Exchange rate

B. U.S. Dollars to Foreign Currency:

Step 1. Locate the exchange rate listed in Column 4, *currency per U.S. $*, for the country in question.

Step 2. Multiply the number of dollars by the exchange rate for that currency:

Foreign currency = U.S. dollars × Exchange rate

EXAMPLE

A Dallas, Texas, importer received an order of merchandise from Mexico with a purchase price of 75,000 pesos. Using Tuesday's exchange rate, how many U.S. dollars will be needed to pay the bill?

SOLUTION STRATEGY

Pesos to U.S. dollars: From Column 2 of the table, we find the exchange rate for Tuesday to be .2994012 dollars per peso.

$$\text{U.S. dollars} = \text{Units of foreign currency} \times \text{Exchange rate}$$

$$\text{U.S. dollars} = 75{,}000 \times .2994012 = \underline{\underline{\$22{,}455.09}}$$

EXAMPLE

A New Jersey exporter sold goods valued at $100,000 to a company in France. The transaction will take place in French currency, using a 30-day forward contract. Use Monday's exchange rate to calculate the price quotation in francs.

SOLUTION STRATEGY

U.S. dollars to French francs—30-days forward: From Column 5 of the table, we find the 30-day forward exchange rate for Monday to be 5.6735 francs per dollar.

$$\text{French francs} = \text{U.S. dollars} \times \text{Exchange rate}$$

$$\text{French francs} = 100{,}000 \times 5.6735 = \underline{\underline{\text{567,350 francs in 30 days}}}$$

TRY-IT EXERCISE

27. You are the sales manager of a company that sells motor parts in many countries. You are going on a selling trip to Brazil next week. Your airline fare and hotel bill will be charged on company credit cards. Your company has allotted an additional $2,000 for out-of-pocket expenses during your stay. Using Tuesday's listings:

a. How many Brazilian cruzeiro real will you have for the trip?

b. Upon returning to the United States, you still have 25,000 cruzeiro real. At the airport you convert them back to dollars. How many dollars will you receive?

Check your answers with the solutions on page 147.

TRY-IT EXERCISE SOLUTIONS

1. $\frac{\text{feet}}{3} = \text{yards} = \frac{105}{3} = \underline{\underline{35 \text{ yd}}}$

2. $\frac{\text{inches}}{12} = \text{feet} = \frac{156}{12} = \underline{\underline{13 \text{ ft}}}$

3. $\frac{\text{feet}}{5{,}280} = \text{miles} = \frac{35{,}000}{5{,}280} = \underline{\underline{6.6 \text{ miles}}}$

4. $\frac{\text{sq inches}}{144} = \text{sq feet} = \frac{540}{144} = \underline{\underline{3.75 \text{ sq ft}}}$

5. acres × 4,840 = sq yard = 5 × 4,840 = $\underline{\underline{24{,}200 \text{ sq yd}}}$

6. $\frac{\text{acres}}{640} = \text{sq miles} = \frac{5{,}696}{640} = \underline{\underline{8.9 \text{ sq mi}}}$

7. $\frac{\text{fl ounces}}{16} = \text{pints} = \frac{80}{16} = \underline{\underline{5 \text{ pints}}}$

8. pecks × 8 = quarts = 12 × 8 = $\underline{\underline{96 \text{ qt}}}$

9. a. bushels × 4 = pecks = 1 × 4 = 4 pecks
   pecks × 8 = quarts = 4 × 8 = 32 qt
   quarts × 2 = pints = 32 × 2 = $\underline{\underline{64 \text{ pints}}}$ in a bushel

   b. sales = 64 pints × 1.19 = \$76.16
   cost of merchandise − 45.00
   $\underline{\underline{\$31.16}}$ gross profit

10. $\frac{\text{ounces}}{16} = \text{pounds} = \frac{48}{16} = \underline{\underline{3 \text{ lb}}}$

11. pounds × 16 = ounces = 4 × 16 = $\underline{\underline{64 \text{ oz}}}$

12. $\frac{\text{ounces}}{16} = \text{pounds} = \frac{88}{16} = 5\frac{1}{2}$ pounds = 5 lb 8 oz

    John: 173 lb 12 oz
    − 5 lb 8 oz
    $\underline{\underline{168 \text{ lb } 4 \text{ oz}}}$ = Sam's weight

13. $\frac{\text{hours}}{24} = \text{days} = \frac{72}{24} = \underline{\underline{3 \text{ days}}}$

14. minutes × 60 = seconds = 2.5 × 60 = $\underline{\underline{150 \text{ sec}}}$

15. $\frac{\text{months}}{12} = \text{years} = \frac{66}{12} = \underline{\underline{5\frac{1}{2} \text{ years}}}$

16. centimeters to meters = 2 places left
    32 cm = $\underline{\underline{.32 \text{ meters}}}$

17. kilograms to grams = 3 places right
    12.5 kg = $\underline{\underline{12{,}500 \text{ grams}}}$

18. liters to milliliters = 3 places right
    6.7 l = $\underline{\underline{6{,}700 \text{ milliliters}}}$

19. hectares × 10,000 = sq meters = 16.5 × 10,000 = $\underline{\underline{165{,}000 \text{ sq meters}}}$

20. $\frac{\text{ares}}{100} = \text{hectares} = \frac{2{,}690}{100} = \underline{\underline{26.9 \text{ hectares}}}$

21. pounds × .4536 = kilograms = 69 × .4536 = $\underline{\underline{31.3 \text{ kg}}}$

22. sq meters × 10.764 = sq feet = 800 × 10.764 = $\underline{\underline{8{,}611.2 \text{ sq ft}}}$

23. meters × 39.37 = inches = 15 × 39.37 = $\underline{\underline{590.6 \text{ inches}}}$

24. liters × .9081 = quarts (dry) = 30 × .9081 = $\underline{\underline{27.2 \text{ qts}}}$

25. $F = 1.8C + 32$

    $F = 1.8(50) + 32$

    $F = 90 + 32$

    $F = \underline{\underline{122°}}$

26. $C = \frac{5}{9}(F - 32)$

    $C = \frac{5}{9}(140 - 32)$

    $C = \frac{5}{9}(108)$

    $C = \underline{\underline{60°}}$

27. a. Foreign currency = U.S. dollars × Exchange rate
    Foreign currency = 2,000.00 × 1,459.36 = $\underline{\underline{2{,}918{,}720 \text{ cruzeiro real}}}$

    b. U.S. dollars = Foreign currency × Exchange rate
    U.S. dollars = 25,000.00 × .0006852 = $\underline{\underline{\$17.13}}$

Answers to Practice Problems

Answers to Chapter 1 Practice Exercises

1. **a.** \$69
b. Over budget
c. \$5

2.

	Monday	Tuesday	Wednesday	Thursday	Friday	Total Meals
Breakfast	52	65	32	57	72	278
Lunch	98	68	41	39	97	343
Dinner	56	95	72	108	112	443
Late Night	208	148	169	214	303	1,042
Daily Totals	414	376	314	418	584	2,106

3. \$1,486

4. **a.** 25,820 tickets
b. \$326,604 total sales

5. **a.** \$27 average per transaction
b. \$3,105 average per day

6. **a.** \$10,055 total expenses for March
b. \$120,660 total expenses per year

7. \$361,500 total sales

8. \$12,200 profit

9. \$2,220,410 over projection

Answers to Chapter 2 Practice Exercises

1. $132\frac{1}{8}$ pounds

2. \$5,100 more

3. \$80 on Tuesday

4. 560 boxes

5. $15\frac{13}{20}$ acres left

6. 554 bins

7.

noodles	60 ounces
sour cream	7 cups
sugar	$1\frac{2}{3}$ cups
cherries	$18\frac{3}{4}$ ounces
vanilla	$1\frac{7}{8}$ teaspoons
eggs	5 eggs

8. 2,880 GM cars

9. $9\frac{1}{20}$ more hours

10. 675 preferred stamps

11. \$5,800 total price

12. \$$52\frac{3}{8}$ new price

13. \$33 deducted for union dues

14. 23 more chairs

15. 80 pieces of rope

16. $240 sale price

17. $4,000 each share

18. **a.** not safe, needs heavy-duty trailer
    b. $203\frac{3}{4}$ pounds over limit

19. 600 loaves

Answers to Chapter 3 Practice Exercises

1. 48.57 gallons used

2. **a.** $580.53 saved
   c. $2,206.97 total purchase price

3. $363.66 total earnings

4. $27.84 total cost

5. $11.60 per hour

6. **a.** 8 minute miles
   b. 12.5 minutes longer for Bryant

7. **a.** $1,427.63 total purchases
   b. $939.80 total savings

8.

Employee	Hours	Hourly Wage	Total	Deductions	Net Pay
Williams	37.6	$6.80	$255.68	$51.68	$204.00
Johnson	43.22	$8.25	$356.57	$102.64	$253.93
Calder	39.37	$10.47	$412.20	$124.25	$287.95
Sanford	27.7	$17.68	$489.61	$122.63	$366.98

9. **a.** 170,000 total sq. ft.
   b. $4,760 total maintenance cost

10. 29 pounds of nails

11. **a.** no—cannot be shipped in one truckload
    b. 157.8 pounds overweight

12. $1,300.00 cost per acre

13. 543 containers

14. **a.** $19,250 cost of shares
    b. $19,708.23 total cost of transaction

Answers to Chapter 4 Practice Exercises

1. a. and b.

IF TAX DEDUCTIBLE CHECK HERE ☐ $ 1,875.40
3088
May 26 19 xx
TO Walton Group
FOR Supplies

	DOLLARS	CENTS
BAL. FWD.	4,599	36
DEPOSIT		
DEPOSIT		
TOTAL	4,599	36
THIS ITEM	1,875	40
SUB-TOTAL	2,723	96
OTHER DEDUCT. (IF ANY)		
BAL. FWD.	2,723	96

Ullman & Company
390 Cherry Rd.
Boise, Idaho 54948

3088
May 26 19 xx
63-398/670

PAY TO THE ORDER OF The Walton Group $ 1,875 40/100
Eighteen Hundred Seventy-Five and 40/100 DOLLARS

Barnett Bank
037-049
11755 Biscayne Blvd.
North Miami, Florida 33161

FOR Supplies Your Signature

⑆067003985⑆ 3088 630211502 ⑈

c.

Ullman & Company
390 Cherry Rd.
Boise, Idaho 54948

DATE June 4 19 XX
DEPOSITS MAY NOT BE AVAILABLE FOR IMMEDIATE WITHDRAWAL

SIGN HERE IF CASH RECEIVED FROM DEPOSIT

Barnett Bank
037
11755 Biscayne Blvd.
North Miami, Florida 33161

CASH CURRENCY	450	00
CASH COIN	66	75
CHECKS		
TOTAL FROM OTHER SIDE	5,690	11
TOTAL	6,206	86
LESS CASH		
NET DEPOSIT	6,206	86

63-398/670
DEPOSIT TICKET
USE OTHER SIDE FOR ADDITIONAL LISTINGS
TOTAL ITEMS
BE SURE EACH ITEM IS PROPERLY ENDORSED

⑆067003985⑆ 630211502 ⑈

REV. 6/88 CHECKS AND OTHER ITEMS ARE RECEIVED FOR DEPOSIT SUBJECT TO THE PROVISIONS OF THE UNIFORM COMMERCIAL CODE OR ANY APPLICABLE COLLECTION AGREEMENT

d.

IF TAX DEDUCTIBLE CHECK HERE ☐ $ 1,391.88
3089
June 9 19 xx
TO Central Supply
FOR Roofing Materials

	DOLLARS	CENTS
BAL. FWD.	2,723	96
DEPOSIT 6/4	6,206	86
DEPOSIT		
TOTAL	8,930	82
THIS ITEM	1,391	88
SUB-TOTAL	7,538	94
OTHER DEDUCT. (IF ANY)		
BAL. FWD.	7,538	94

Ullman & Company
390 Cherry Rd.
Boise, Idaho 54948

3089
June 9 19 xx
63-398/670

PAY TO THE ORDER OF Central Supply, Inc. $ 1,391 88/100
Thirteen Hundred Ninety-one and 88/100 DOLLARS

Barnett Bank
037-049
11755 Biscayne Blvd.
North Miami, Florida 33161

FOR Roofing Materials Your Signature

⑆067003985⑆ 3089 482130150 ⑈

2. a.
for deposit
Your Signature
265-55-7

Full Endorsement

b.
Your Signature
265-55-7

Blank Endorsement

c.
Pay to the order of
Beverly Baker
Your Signature
265-55-7

Restrictive Endorsement

3.

CHECKBOOK BALANCE	$ 852.80	STATEMENT BALANCE	$ 788.28
Deduct: Service Charges & Other Debits	9.60	Deduct: Outstanding Checks	191.19
SUB TOTAL	843.20	SUB TOTAL	597.09
Add: Interest Earned & Other Credits	—	Add: Deposits in Transit	246.11
ADJUSTED CHECKBOOK BALANCE	843.20	ADJUSTED STATEMENT BALANCE	843.20

Reconciled Balances

Checks Outstanding

No.	Amount	
	42	50
	12	80
	108	17
	27	12
Total	191	19

4.

CHECKBOOK BALANCE	$ 5,489.01	STATEMENT BALANCE	$ 5,595.37
Deduct: Service Charges & Other Debits	575.00	Deduct: Outstanding Checks	1,942.50
SUB TOTAL	4,914.01	SUB TOTAL	3,652.87
Add: Interest Earned & Other Credits	2,000.00	Add: Deposits in Transit	3,652.14
ADJUSTED CHECKBOOK BALANCE	6,914.01	ADJUSTED STATEMENT BALANCE	6,914.01

Reconciled Balances

Checks Outstanding

No.	Amount	
	262	07
	467	50
	1,212	93
Total	1,942	50

5.

CHECKBOOK BALANCE	$ 543.46	STATEMENT BALANCE	$ 1,106.06
Deduct: Service Charges & Other Debits	143.00	Deduct: Outstanding Checks	505.60
SUB TOTAL	400.46	SUB TOTAL	600.46
Add: Interest Earned & Other Credits	300.00	Add: Deposits in Transit	100.00
ADJUSTED CHECKBOOK BALANCE	700.46	ADJUSTED STATEMENT BALANCE	700.46

Reconciled Balances

Checks Outstanding		
No.	Amount	
502	319	43
504	186	17
Total	505	60

Answers to Chapter 5 Practice Exercises

1. Tony $= X$
George $= X - 6$
$X + X - 6 = 46$

$X = 26 =$ Tony's sales
$X - 6 = 20 =$ George's sales

2. $X =$ total trucks
$\frac{1}{6}X = 540$
$X = 3{,}240$ trucks

3. $X =$ cost per stockholder
$244X = 6{,}222$
$X = \$25.50 =$ cost per stockholder

4. men's shoes $= X$
women's shoes $= 3\frac{1}{2}X$
$X + 3\frac{1}{2}X = 65{,}430$

$X = \$14{,}540 =$ men's shoe sales
$3\frac{1}{2}X = \$50{,}890 =$ women's shoe sales

5. Los Angeles plant $= X$
San Francisco $= 2X + 200$
$2X + 200 = 1500$
$X = 650 =$ units per hour, L.A. plant

6. **a.** large frame $= X$
small frames $= 88 - X$
$7.50X + 4.50(88 - X) = 471$

$X = 25 =$ large frames
$88 - X = 63 =$ small frames

b. \$187.50 large frame sales
\$283.50 small frame sales

7. Lucas' investment $= X$
Brown's investment $= 4X$
$X + 4X = \$128{,}000$

$X = \$25,600$ = Lucas' investment
$4X = \$102,400$ = Brown's investment

8. Jim $= X$
Fred $= X$
Lisa $= 3X$
Annie $= 4X$
$X + X + 3X + 4X = \$376,200$

Jim's share $= X = \$41,800$
Fred's share $= X = \$41,800$
Lisa's share $= 3X = \$125,400$
Annie's share $= 4X = \$167,200$

9. total calculator sales $= X$
solar powered sales $= \frac{1}{4}X - 14$
$\frac{1}{4}X - 14 = 44$

$X = 232$ = total calculator sales

10. cordless phones $= X$
cellular phones $= 3X + 35$
$X + 3X + 35 = 549$

$X = \$128.50$ = cost of cordless phone
$3X + 35 = \$420.50$ = cost of cellular phone

11. **a.** leather sofas $= X$
fabric sofas $= 36 - X$
$900X + 650(36 - X) = \$28,400$

$X = 20$ = leather sofas
$36 - X = 16$ = fabric sofas

b. leather $= \$18,000$
fabric $= \$10,400$

12. Blakley Building $= X$
Winston Building $= 3X$
$3X - X = 88$

$X = 44$ = height of Blakley Building
$3X = 132$ = height of Winston Building

13. Total employees $= X$
$\frac{1}{20}X$ = employees with over 20 years of service
$\frac{1}{20}X = 42$

$X = 840$ = total employees

14. scallops $= X$
shrimp $= 2X$
$6.50X + 8.50(2X) = \$1,316$

$X = 56$ = pounds of scallops
$2X = 112$ = pounds of shrimp

15. X = price per piece
$8(5X) = 480$
$X = \$12.00$ = price per piece

16. X = total cost to ship order
$X = .15(350) + .08(350) + .27(350) + 42.50$

$X = \$217.50$ total cost

17. \$344.72 interest

18. 200 hot dogs sold

19. \$5,600 disability coverage

20. 3 teaspoons of vanilla

21. 176 miles
22. 1,210 miles
23. 240 sweaters
24. 1,050 children

Answers to Chapter 6 Practice Exercises

1. 70.6%
2. $87,675
3. 17%
4. 1,700 people
5. 172 employees
6. 8.3%
7. $1,550,000 total sales
8. Ethel's share = $33,750
   Goldie's share = $46,250
9. $12,000 total sales
10. 36.2%
11. $2,255 September sales
12. **a.** 23% for deductions
    b. $462 new take-home pay
13. 15.6% increase
14. **a.** 85%
    b. $12,320 original price
15. **a.** 3,800,000 state population
    b. 266,000 population 10 years ago
    c. 799,330 projected population in 5 years
16. **a.** $6.40 amount profit will go up
    b. 11.7% increase
17. $1,250 fruit sales

Answers to Chapter 7 Practice Exercises

1. $375.00
2. **a.** $786.00
   b. $1,834.00
3. $1,834.00
4. 23.68%
5. Premium Paper Co. offers lower net price.
6.

Quantity	**Extended Total**
250	$13,762.35
400	$17,546.87
124	$19,048.88
Invoice Subtotal	$50,358.10
$4\frac{1}{2}$% Volume Discount	− 2,266.11
Invoice Total	$48,091.99

7. Net price = $5,520.00
   Trade discount = $4,480.00

8. Net price factor = .5984
   Net price = $777.92

9. Net price factor = .56525
   Single equivalent discount = .43475

10. Net price factor = .64125
    Single equivalent discount = .35875

11. **a.** If the balance is paid within 15 days, an extra 3% cash discount may be taken. To avoid penalty, the invoice must be paid within 30 days of the invoice date.
    b. Paid in 20 days, take additional 2% cash discount. Net amount within 60 days.
    c. Paid in 20 days, take additional 4% cash discount; within 30 days, take a 2% cash discount. Net amount due within 60 days.
    d. Pay within 15 days after receiving goods, take 3% cash discount. Net amount due within 45 days of receipt of goods.
    e. If invoice is dated 1st through 25th, an additional 4% cash discount can be taken if paid by 15 days after the end of the month of the invoice. If the invoice is dated after the 25th, an additional 4% cash discount can be taken if paid by 15 days after the end of the following month. In both cases, the net date is 20 days after the discount date.
    f. A 2% cash discount can be taken if paid within 55 days (15 + 40 extra). The net date is 20 days after the discount date.

12. **a.** Discount date = Jan 30
    Net date = Feb 14
    b. Discount date = Feb 19
    Net date = Mar 11

13. **a.** Discount date = July 18
    Net date = Aug 12
    b. Partial payment credit = $1,250.00
    c. Balance due = $1,500.00

Answers to Chapter 8 Practice Exercises

1. $42.80

2. $4.12

3. $47.70

4. **a.** $240.30
   b. 55.2%

5. **a.** $11.45
   b. 70.9%

6. **a.** $11.55
   b. 36.7%
   c. 57.9%

7. **a.** $21,725.00
   b. 45%
   c. 81.8%

8. $133.33

9. $58.78

10. **a.** 37%
    b. 75%

11. **a.** $15.55
    b. 34.2%

12. **a.** $351.00
    b. 20%

13. $37.45

14. $299.55

15. $144.08

16. $1.25

17. **a.** bench: $80.00
    b. weights: $16.00

18. $109.36

19. **a.** $129.97
    b. $159.97

Answers to Chapter 9 Practice Exercises

1. $2,870.83

2. $378.00

3. $781.25

4. $472.50

	Employee	Total Hours	Overtime Hours	Regular Pay	Overtime Pay	Total Pay
5.	Proctor	38	0	$256.50	0	$256.50
6.	Wilson	53	13	$360.00	$175.50	$535.50
7.	Chalks	44	4	$548.00	$82.20	$630.20

8. $452.52

9. $324.35

	Employee	Total Gross Pay
10.	Peterson, B.	$212.16
11.	Jiminez, L.	$416.85
12.	Morgan, S.	$848.25

13. $7,803.00

14. $7,761.90

15. $2,089.50

16. $4,792.00

17. $435.30

18. **a.** $39,000.00
    b. $17,550.00

19. Social Security: $44.95 per week
    Medicare: $10.51 per week

20. Social Security: $103.11 semi-monthly
    Medicare: $24.11 semi-monthly

21. **a.** Social Security: $545.60 for January Medicare: $127.60 for January
    b. August (7 full months plus a portion of the 8th month)
    c. Social Security: $68.20 for August Medicare: $127.60 for August

	Employee	Social Security	Medicare
22.	Kloot	$179.80	$42.05
23.	Jefferson	$179.18	$59.45
24.	Rodriguez	0	$161.68

25. $151.00

26. $545.00

27. $2,053.57 net pay per month

	Employee	Income Tax Withholding
28.	Alders	$788.00
29.	Nunez	$91.00
30.	Burkett	$129.00
31.	Wong	$370.00

	Employee	Income Tax Withholding
32.	Rockwell	$55.18
33.	Kelman	$200.89
34.	Scher	$1,040.62
35.	Segall	$174.11

36. $3,810.27 net pay

37. **a.** Social Security: $1,141.05 per week   Medicare: $266.86 per week
    b. Social Security: $14,833.65 per quarter   Medicare: $3,469.18 per quarter

38. Social Security: $773.76   Medicare: $180.96

	Employee	Social Security	Medicare	Income Tax	Net Pay
39.	Mann	$31.99	$7.48	$73.41	$315.12
40.	Tolbert	$233.24	$54.55	$713.23	$2,713.98
41.	Eason	$223.20	$57.03	$932.04	$2,546.73
42.	Best	0	$220.40	$3,596.42	$11,119.18

43. **a.** $378.00 SUTA per year
    b. $56.00 FUTA per year

44. **a.** SUTA 1st quarter = $953.64   FUTA 1st quarter = $141.28
    b. SUTA 2nd quarter = $180.36   FUTA 2nd quarter = $26.72

45. **a.** $42,402.50
    b. 16.7%
    c. $2,204,930.00

46. **a.** $4,570.45
    b. 1040-ES Estimated Quarterly Tax Voucher for Self-Employed Persons

47. **a.** Monthly
    b. Form 8109
    c. Quarterly
    d. Form 941 Employer's Quarterly Federal Tax Return

Answers to Chapter 10 Practice Exercises

1. **a.** $29.33
   b. $829.33
   c. December 31

2. **a.** $11,000.00
   b. July 12

3. $158.56

4. $12,000.00

5. 9.9%

6. **a.** 105 days
   b. February 25

7. 5 years

8. $7,223.84

9. **a.** $46,097.49

b. August 14

10. a. $34,115.28
    b. August 28

11. 12.91%

12. a. May 7
    b. $288,594.44
    c. April 12
    d. $285,387.84

Answers to Chapter 11 Practice Exercises

1. $53,875.80

2. Yes—$155,421.50

3. $18,708.60

4. $3,935.55

5. 12.68%

6. a. $40,722.50
   b. $15,722.50
   c. 10.25%

7. $25,236.38

8. a. 8.24%
   b. $15,456.35
   c. $3,116.75

9. a. $5,153.58
   b. $1,646.42

10. $77,419.09

11. a. $37,243.34
    b. $14,243.34

12. $556.90 difference

Answers to Chapter 12 Practice Exercises

1. $11,807.80

2. $5,453.76

3. $10,543.38

4. a. $53,034.25
   b. $17,034.25
   c. $106,068.51

5. $388,965.00

6. a. $783.04
   b. $759.66

7. a. $149.77
   b. $8.55 per month

8. a. $125,650.79
   b. $492,190.60

9. $7,667.74

10. $41,002.00

11. **a.** $50,632.91
b. $40,123.87

12. $34,885.02

13. $158,773.77

14. $25,102.78

15. $5,198.42

16. $39,775.07

17. $1,131.20

Answers to Chapter 13 Practice Exercises

1. **a.** $5.76
b. $741.62

	Month	Previous Month's Balance	Finance Charge	New Balance End of Month
2.	August	$740.00	$11.10	$844.76
3.	September	$844.76	$12.67	$838.91
4.	October	$838.91	$12.58	$951.84
5.	November	$951.84	$14.28	$1,018.77

6. $631.85

7. **a.** $5.83
b. $840.84

8. **a.** 10% minimum
b. 13.35%

9. **a.** $539.42
b. $26,339.42
c. $48,660.58

10. $8,953.75

11. **a.** $3,068.65
b. $12,967.65

12. **a.** $1,034.30
b. $2,694.30

13. **a.** $4,557.07
b. $81.31

14. **a.** $14,690.00
b. $306.04

15. **a.** $4,480.04
b. 16.25%

16. **a.** 20.26%
b. 20.75%

17. **a.** $3,394.63
b. $481.55

18. **a.** 378
b. 666
c. 105
d. 1,830

19. **a.** 300
b. 2,628

20. **a.** $10,710.00
b. $879.38
c. 15.24%
d. $13,556.07

Answers to Chapter 14 Practice Exercises

	Number of $1,000s Financed	Table Factor	Monthly Payment	Total Interest
1.	99	7.34	$726.66	$162,597.60
2.	52.8	7.07	373.30	59,190.00
3.	170.7	10.45	1,783.82	150,387.60
4.	229.6	9.34	2,144.46	542,405.60
5.	1,000	9.71	9,710.00	747,800.00

6.

	Monthly Payment	Monthly Interest	Portion Used to Reduce Principal	Loan Balance
0				$122,000.00
1	$983.32	$864.17	119.15	121,880.85
2	983.32	863.32	120.00	121,760.85
3	983.32	862.47	120.85	121,640.00
4	983.32	861.62	121.70	121,518.30

7. a. 8.25%
   b. 15.25%

8. a. $500.00
   b. $26.80

9. a. $604.10 PI
   b.

	Monthly Payment	Monthly Interest	Portion Used to Reduce Principal	Loan Balance
0				$58,650.00
1	$604.10	$452.09	$152.01	58,497.99
2	604.10	450.92	153.18	58,344.81
3	604.10	449.74	154.36	58,190.45
4	604.10	448.55	155.55	58,034.90
5	604.10	447.35	156.75	57,878.15

   c. $766.60 PITI

10. $3,174.60

11. a. Bentley Bank: $2,724,000.00 total interest
       Silver Savings: $2,805,000.00 total interest
    b. Bentley Bank is a better deal by $13,500.

12. a. 20-year mortgage: $559.44 monthly PI
       25-year mortgage: $569.52 monthly PI
    b. 20-year mortgage: $78,265.60 total interest
       25-year mortgage: $114,856.00 total interest
    c. $24,494.40 more interest paid

13. $43,310.00

14. $256,000.00

15. a. $844.86 PITI
    b. $145,149.60 total interest paid
    c. $10,531.26 due at closing
    d. $56,065.00 seller's proceeds

16. a. 20% housing expense ratio
    b. 28.59% total obligations ratio
    c. She qualifies for FHA and Conventional loans.

Answers to Chapter 15 Practice Exercises

1. a.

Party-Time Rentals
Balance Sheet
December 31, 1996

Assets		**Percent**
Current Assets	$232,540	71.1
Property, Plant, and Equipment	65,430	20.0
Investments and Other Assets	29,000	8.9
Total Assets	326,970	100.0
Liabilities		
Current Liabilities	69,600	21.3
Long-term Liabilities	167,400	51.2
Total Liabilities	237,000	72.5
Owner's Equity	89,970	27.5
Total Liabilities and Owner's Equity	326,970	100.0

b. **Party-Time Rentals**
Comparative Balance Sheet
December 31, 1996 and 1997

Assets	1997	1996	Amount	Percent
Current Assets	$265,300	$232,540	$32,760	14.1
Property, Plant, and Equipment	71,200	65,430	5,770	8.8
Investments and Other Assets	29,000	29,000	0	0
Total Assets	365,500	326,970	38,530	11.8
Liability and Owners Equity				
Current Liabilities	51,500	69,600	(18,100)	(260)
Long-Term Liabilities	189,800	167,400	22,400	13.4
Total Liabilities	241,300	237,000	4,300	1.8
Owners Equity	124,200	89,970	34,230	38.0
Total Liabilities and Owner's Equity	365,500	326,970	38,530	11.8

2. a. **Dollar Days Department Store, Inc.**
Balance Sheet
April 30, 1996

Assets

Current Assets		Percent
Cash	$ 56,700	2.9
Accounts Receivable	314,300	15.9
Merchandise Inventory	248,500	12.5
Prepaid Insurance	6,900	.3
Maintenance Supplies	2,330	.1
Total Current Assets	628,730	31.7
Property, Plant, and Equipment		
Land	214,400	10.8
Buildings	717,000	36.2
Fixtures	101,900	5.1
Delivery Trucks	73,900	3.7
Computers	43,700	2.2
Total Property, Plant, and Equipment	1,150,900	58.0
Investments and Other Assets		
Investments	122,400	6.2
Goodwill	80,000	4.0
Total Assets	$1,982,030	100%

Liabilities and Owner's Equity

Current Liabilities		
Accounts Payable	85,900	4.3
Salaries Payable	38,200	1.9
Notes Payable	152,000	7.7
Total Current Liabilities	276,100	13.9
Long-Term Liabilities		
Mortgage Payable	494,800	25.0
Debenture Bonds	138,000	7.0
Total Long-Term Liabilities	632,800	32.0
Total Liabilities	908,900	45.9
Owner's Equity		
Common Stock	875,000	44.1
Retained Earnings	198,130	10.0
Total Owner's Equity	1,073,130	54.1
Total Liabilities and Owner's Equity	$1,982,030	100%

Note: Percents may vary by .1 due to rounding

b. Dollar Days Department Stores, Inc.
Comparative Balance Sheet
April 30, 1996 and 1997

Assets	1997	1996	Increase/Decrease Amount	Percent
Current Assets				
Cash	$ 65,400	$ 56,700	8,700	15.3
Accounts Receivable	365,420	314,300	51,120	16.3
Merchandise Inventory	213,700	248,500	(34,800)	(14.0)
Prepaid Insurance	7,300	6,900	400	5.8
Maintenance Supplies	3,420	2,330	1,090	46.8
Total Current Assets	655,240	628,730	26,510	4.2
Property, Plant, and Equipment				
Land	214,400	214,400	0	0
Buildings	815,000	717,000	98,000	13.7
Fixtures	132,760	101,900	30,860	30.3
Delivery Trucks	68,700	73,900	(5,200)	(7.0)
Computers	42,100	43,700	(1,600)	(3.7)
Total Property, Plant, and Equipment	1,272,960	1,150,900	122,060	10.6
Investments and Other Assets				
Investments	145,600	122,400	23,200	19.0
Goodwill	90,000	80,000	10,000	12.5
Total Assets	$2,163,800	$1,982,030	181,770	9.2
Liabilities and Owner's Equity				
Current Liabilities				
Accounts Payable	96,500	85,900	10,600	12.3
Salaries Payable	33,600	38,200	(4,600)	(11.9)
Notes Payable	143,120	152,000	(8,880)	(5.8)
Total Current Liabilities	273,220	276,100	(2,800)	(1.0)
Long-Term Liabilities				
Mortgage Payable	481,300	494,800	(13,500)	(2.7)
Debenture Bonds	138,000	138,000	0	0
Total Long-Term Liabilities	619,300	632,800	(13,500)	(2.1)
Total Liabilities	892,520	908,900	(16,380)	(1.8)
Owner's Equity				
Common Stock	875,000	875,000	0	0
Retained Earnings	396,280	198,130	198,150	100.0
Total Owner's Equity	1,271,280	1,073,130	198,150	18.5
Total Liabilities and Owner's Equity	$2,163,800	$1,982,030	181,770	9.2

3. $346,376 net sales

4. **a.** $1,878,760 goods available for sale
b. $1,019,920 cost of goods sold
c. $630,080 gross margin
d. $242,780 net profit

5. a. **American Eagle Paper Company**
Income Statement
January 1–March 31, 1997

Revenue		
Gross Sales	$748,000	123.3
Less: Sales Returns and Allowances	141,200	23.3
Net Sales	606,800	100.0
Cost of Goods Sold		
Merchandise Inventory, Jan. 1	457,400	75.5
Net Purchases	279,600	46.1
Goods Available for Sale	737,000	121.5
Less: Merchandise Inventory, March 31	517,300	85.3
Cost of Goods Sold	219,700	36.2
Gross Margin	387,100	63.8
Operating Expenses	328,000	54.1
Income before Taxes	59,100	9.7
Income Tax	16,530	2.7
Net Income	$ 42,570	7.0

b. **American Eagle Paper Company**
Comparative Income Statement
First and Second Quarter, 1997

			Increase/Decrease	
	2nd Qtr	**1st Qtr**	**Amount**	**Percent**
Revenue				
Gross Sales	$784,330	$748,000	36,330	4.9
Less: Sales Returns and Allowances	167,400	141,200	26,200	18.6
Net Sales	616,930	606,800	10,130	1.7
Cost of Goods Sold				
Merchandise Inventory, Beginning	425,430	457,400	(31,970)	(7.0)
Net Purchases	291,700	279,600	12,100	4.3
Goods Available for Sale	717,130	737,000	(19,870)	(2.7)
Less: Merchandise Inventory, Ending	486,800	517,300	(30,500)	(5.9)
Cost of Goods Sold	230,330	219,700	10,630	4.8
Gross Margin	386,600	387,100	(500)	(.1)
Operating Expenses	292,710	328,000	(35,290)	(10.8)
Income before Income Tax	93,890	59,100	34,790	58.9
Income Tax	36,200	16,530	19,670	119.0
Net Income	$ 57,690	$ 42,570	15,120	35.5

6. a. **Shamrock Sporting Goods, Inc.**
Income Statement
For the year ended December 31, 1996

Revenue		
Gross Sales	$3,010,000	112.0
Less: Sales Returns and Allowances	254,700	9.5
Sales Discounts	67,190	2.5
Net Sales	$2,688,110	100.0
Cost of Goods Sold		
Merchandise Inventory, Jan. 1	1,344,900	50.0
Net Purchases	634,880	23.6
Freight In	21,320	.8
Goods Available for Sale	2,001,100	74.4
Less: Merchandise Inventory, Dec. 31	916,110	34.1
Cost of Goods Sold	1,084,990	40.4
Gross Margin	1,603,120	59.6

Operating Expenses		
Salaries	122,300	4.5
Commissions	379,900	14.1
Rent	285,100	10.6
Depreciation	77,300	2.9
Utilities	15,650	.6
Advertising	441,690	16.4
Insurance	27,710	1.0
Office Expenses	91,480	3.4
Miscellaneous Expenses	88,300	3.3
Total Operating Expenses	1,529,430	57.0
Income before Taxes	73,690	2.7
Income Tax	24,290	.9
Net Income	$ 49,400	1.8

b.

Shamrock Sporting Goods, Inc.
Comparative Income Statement
For the Years ended December 31, 1996 and 1997

			Increase/Decrease	
	1997	1996	Amount	Percent
Revenue				
Gross Sales	$3,129,800	$3,010,000	119,800	4.0
Less: Sales Returns and Allowances	245,400	254,700	(9,300)	(3.7)
Sales Discounts	88,940	67,190	21,750	32.4
Net Sales	2,795,460	2,688,110	107,350	4.0
Cost of Goods Sold				
Merchandise Inventory, Jan. 1	1,698,800	1,344,900	353,900	26.3
Net Purchases	723,230	634,880	88,350	13.9
Freight In	31,210	21,320	9,890	46.4
Goods Available for Sale	2,453,240	2,001,100	452,140	22.6
Less: Merchandise Inventory, Dec. 31	1,307,690	916,110	391,580	42.7
Cost of Goods Sold	1,145,550	1,084,990	60,560	5.6
Gross Margin	1,649,910	1,603,120	46,790	2.9
Operating Expenses				
Salaries	103,600	122,300	(18,700)	(15.3)
Commissions	411,120	379,900	31,220	8.2
Rent	292,190	285,100	7,090	2.5
Depreciation	60,050	77,300	(17,250)	(22.3)
Utilities	17,530	15,650	1,880	12.0
Advertising	440,900	441,690	(710)	(.2)
Insurance	25,200	27,710	(2,510)	(9.1)
Office Expenses	94,150	91,480	2,670	2.9
Miscellaneous Expenses	82,500	88,300	(5,800)	(6.6)
Total Operating Expenses	1,527,240	1,529,430	(2,190)	(.1)
Income before Income Tax	122,670	73,690	48,980	66.5
Income Tax	33,590	24,290	9,300	38.3
Net Income	$ 89,080	$ 49,400	39,680	80.3

	Company	Working Capital	Current Ratio
7.	Quality Copiers, Inc	$224,520	1.7:1
8.	Anchor Fence Company	$46,250	1.61:1
9.	Elite Beauty Supply	$15,447	1.25:1

	Company	Quick Assets	Acid Test Ratio
10.	Quality Copiers, Inc	$307,890	.96:1
11.	Anchor Fence Company	$89,630	1.19:1
12.	Elite Beauty Supply	$56,710	.93:1

13. 48 days

14. a. 65 days
    b. change was bad
    c. Elite takes 25 days longer to collect bills—needs improvement

	Company	Average Inventory	Inventory Turnover
15.	Worldwide Electric	$208,527.50	7.87
16.	Laurenzo's Market	$532,709.00	12.09
17.	Diversified Industries	$11,174,848.00	1.91

18. .65:1

	Company	Owner's Equity	Debt-to-Assets Ratio	Debt-to-Equity Ratio
19.	Cellular Warehouse	$310,020	63.8%	176.3%
20.	Delta Investments	$1,044,850	67.4%	207.1%
21.	Budget Hardware	$91,440	30.3%	43.5%

	Company	Gross Profit	Net Profit	Gross Profit Margin	Net Profit Margin
22.	Skyline Roofing	$493,600	$120,140	.54	.13
23.	Hydraulic Sales	$396,910	$214,110	.17	.09
24.	High-Tech Elec.	$160,175	($17,225)	.35	0

	Company	Return on Investment
25.	Skyline Roofing	22.2%
26.	Hydraulic Sales	26.0%
27.	High-Tech Elec.	0

28.		1996	1995	1994	1993
	Net Sales	132.7%	130.8%	73.0%	100.0%
	Net Income	134.1%	123.0%	16.4%	100.0%
	Total Assets	110.7%	107.0%	96.0%	100.0%
	Stockholder's Equity	107.0%	104.1%	83.7%	100.0%

Answers to Chapter 16 Practice Exercises

1. Total units available: 1,204
   Cost of goods available for sale: $7,701.77

2. a. $1,682.20
   b. $1,690.00
   c. $1,664.00

3. a. 600
   b. $99,827.00

c. $28,073.10
d. $27,891.35
e. $27,785.46

4.

Description	Valuation Basis	Amount
Color monitor	Cost	$20,187.20
Fax modem	Cost	$9,550.60
Hard drive	Market	$3,451.76
Hard drive	Market	$30,447.44
RAM chip	Market	$28,060.00
RAM chip	Cost	$37,051.20
	Total value of inventory	$138,748.20

5. $208,600.00 ending inventory at cost

6. $335,908 estimated ending inventory

7. $154,080 ending inventory at cost

8. $620,652 estimated ending inventory

9. a. $58,400
   b. 4.2 times

10. a. $76,340
    b. 4.7 times

11. a. $432,900 average inventory 9.0 times inventory turnover
    b. $324,166.67 target inventory

12. a. $87,676.67 average inventory 4.7 times inventory turnover
    b. $66,935.48 target inventory

Answers to Chapter 17 Practice Exercises

1.

End of Year	Annual Depreciation	Accumulated Depreciation	Book Value
			(new) $444,775
1	$61,682.14	$61,682.14	383,092.86
2	61,682.14	123,364.28	321,410.72
3	61,682.14	185,046.42	259,728.58
4	61,682.14	246,728.56	198,046.44
5	61,682.14	308,410.70	136,364.30
6	61,682.14	370,092.84	74,682.16
7	61,682.16*	431,775.00	13,000.00

*due to rounding

2.

End of Year	Total Depreciation	Depreciation Rate Fraction	Annual Depreciation	Accumulated Depreciation	Book Value
					(new) $85,000.00
1	$80,000.00	5/15	$26,666.67	$26,666.67	58,333.33
2	80,000.00	4/15	21,333.33	48,000.00	37,000.00
3	80,000.00	3/15	16,000.00	64,000.00	21,000.00
4	80,000.00	2/15	10,666.67	74,666.67	10,333.33
5	80,000.00	1/15	5,333.33	80,000.00	5,000.00

3.

End of Year	Beginning Book Value	Depreciation Rate	Depreciation for the Year	Accumulated Depreciation	Ending Book Value
					(new) $23,500.00
1	$23,500.00	28.57%	$6,713.95	$ 6,713.95	16,786.05
2	16,786.05	28.57%	4,795.77	11,509.72	11,990.28
3	11,990.28	28.57%	3,425.62	14,935.34	8,564.66
4	8,564.66	28.57%	2,446.92	17,382.26	6,117.74
5	6,117.74	28.57%	1,747.84	19,130.10	4,369.90
6	4,369.90	28.57%	1,248.48	20,378.58	3,121.42
7	3,121.42	28.57%	821.42*	21,200.00	2,300.00

*maximum allowed to reach salvage value

4.

End of Year	Depreciation per Hour	Annual Depreciation	Accumulated Depreciation	Book Value
				(new) $628,500.00
1	30.425	$136,912.50	$136,912.50	491,587.50
2	30.425	187,113.75	324,026.25	304,473.75
3	30.425	176,465.00	500,491.25	128,008.75
4	30.425	108,008.75	608,500.00	20,000.00

5. a. $4,216.67 per year
   b. $7,228.57 first year
   c. $8,833.33 first year; largest write-off, double-declining balance
   d. $4,975.67 first year

6. $29,280

7. a. 3 year depreciation
   b. 44.45%
   c. $629,850

8.

End of Year	Original Cost	Cost Recovery Percentage	Cost Recovery (depreciation)	Accumulated Depreciation	Book Value
					(new) $970,000.00
1	$970,000	20%	$194,000.00	$194,000.00	776,000.00
2	970,000	32%	310,400.00	504,400.00	465,600.00
3	970,000	19.20%	186,240.00	690,640.00	279,360.00
4	970,000	11.52%	111,744.00	802,384.00	167,616.00
5	970,000	11.52%	111,744.00	914,128.00	55,872.00
6	970,000	5.76%	55,872.00	970,000.00	0

9. a. $1.596 per barrel
   b. $215,460.00 first year

Answers to Chapter 18 Practice Exercises

1. a. $10.37
   b. $164.47

2. a. $215.00
   b. $12.90

3. a. $44,500.00
   b. $3,337.50

4. a. $56,100
   b. $60,000
   c. $966,100

5. $3,850

6. a. $3,154,800
   b. $39,024.88

7. **a.** $2,364.39
   b. $70.93
   c. $2,541.72

8. **a.** .09%
   b. $.09 per $100
   c. $.90 per $1,000
   d. .9 mills

9. $4,819.00

10. $16,821.00

11. $19,182.00

12. $3,821.00

13. $63,328.54

14. $28,669.00

15. $53,109.22

16. $77,904.42

17. $347

18. $624

19. $101

20. $888

21. **a.** $97,250
    b. $71,210
    c. $17,038
    d. $838

22. Tax liability: $11,821.25 Net income after taxes: $55,463.75

23. Tax liability: $95,063.00 Net income after taxes: $191,637.00

24. Tax liability: $6,252,000.00 Net income after taxes: $11,648,000.00

Answers to Chapter 19 Practice Exercises

1. **a.** $138.50
   b. $72.02

2. **a.** $35.70
   b. $3.21
   c. $2.82 more

3. **a.** cash value = $63,125.00
   b. reduced paid up = $125,000
   c. extended term = Life

4. $232,000

5. $6,000 more life insurance with whole life

6. $27,115 total annual premium

7. $645.00 short-rate refund

8. $145.83 regular refund

9. $61,750.00

10. **a.** $92,000.00
    b. $74,000.00
    c. $34,000.00

11. $736.00

12. $505.40

13. $1,043.00

14. $317.00

15. $1,600.00

16. $248.20 total annual premium

17. $5,590.00

18. a. Insurance will pay:

Steward	$15,000.00
Johnson	8,000.00
Garfield	7,000.00
Cadillac	7,300.00
Gordon's car	2,920.00
	$40,220.00

b. Gordon will pay:

Steward	$5,000.00	
Johnson	0	
Garfield	4,000.00	($30,000 limit reached)
Cadillac	0	
Gordon's car	500.00	(deductible)
	$9,500.00	

Answers to Chapter 20 Practice Exercises

1. $.57 per share

2. a. $3,920,000.00
   b. $3.43 per share

3. a. $12,825,000.00 ($13.50 per share)
   b. $1.89 per share

4. a. 2.3%
   b. 12

5. a. $31,889.14
   b. $34,498.41
   c. $2,609.27

6. a. 8%
   b. $15,910.95
   c. $15,107.85

7. a. $.45
   b. 3.6%

8. $6.44

9. 2,158.273

10. 46.6%

Answers to Chapter 21 Practice Exercises

1. a.

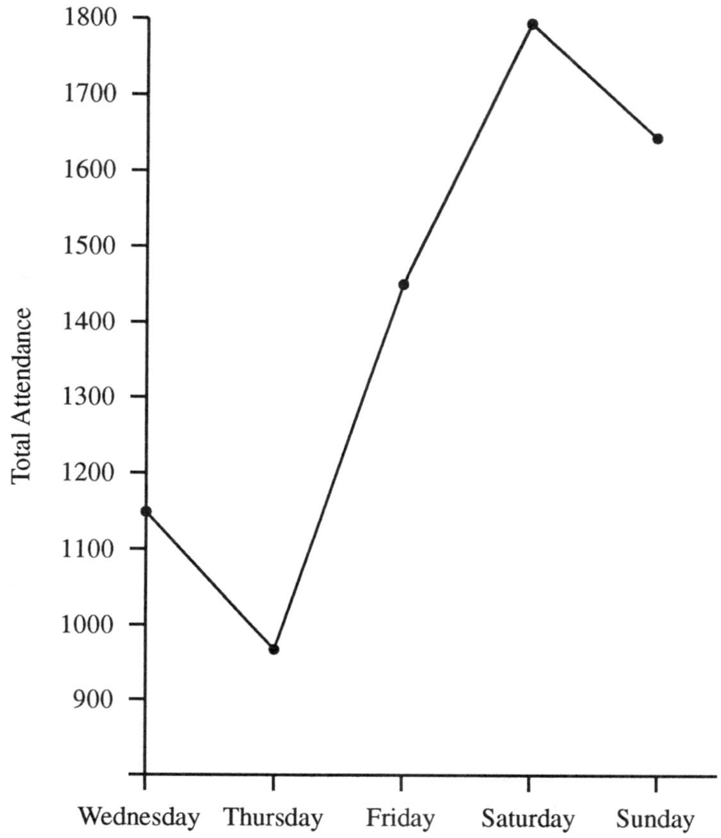

b.

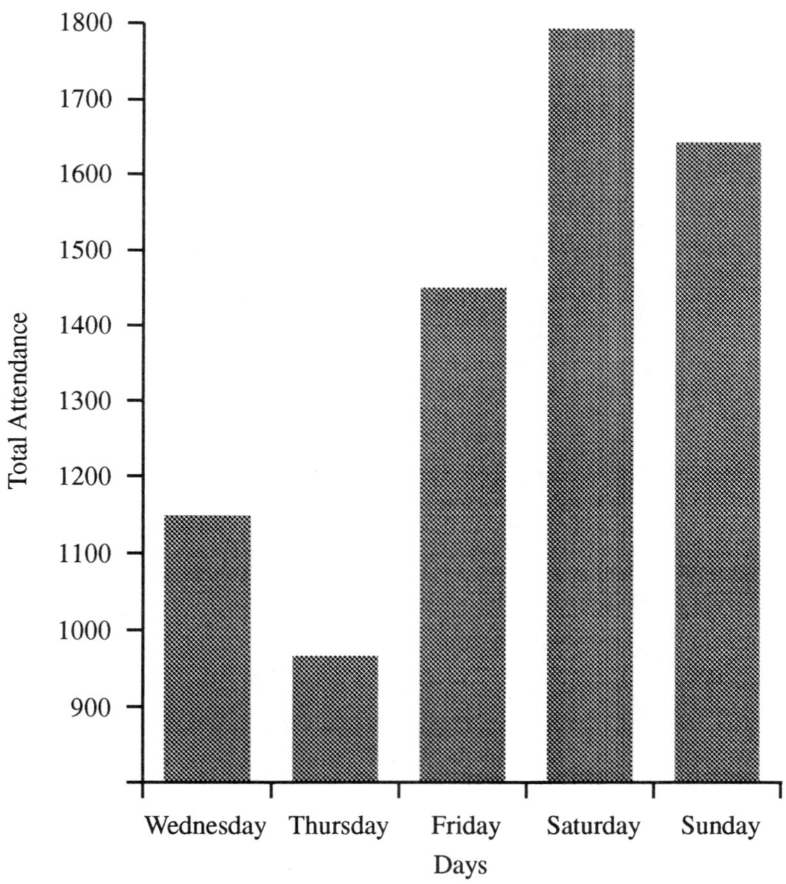

c.

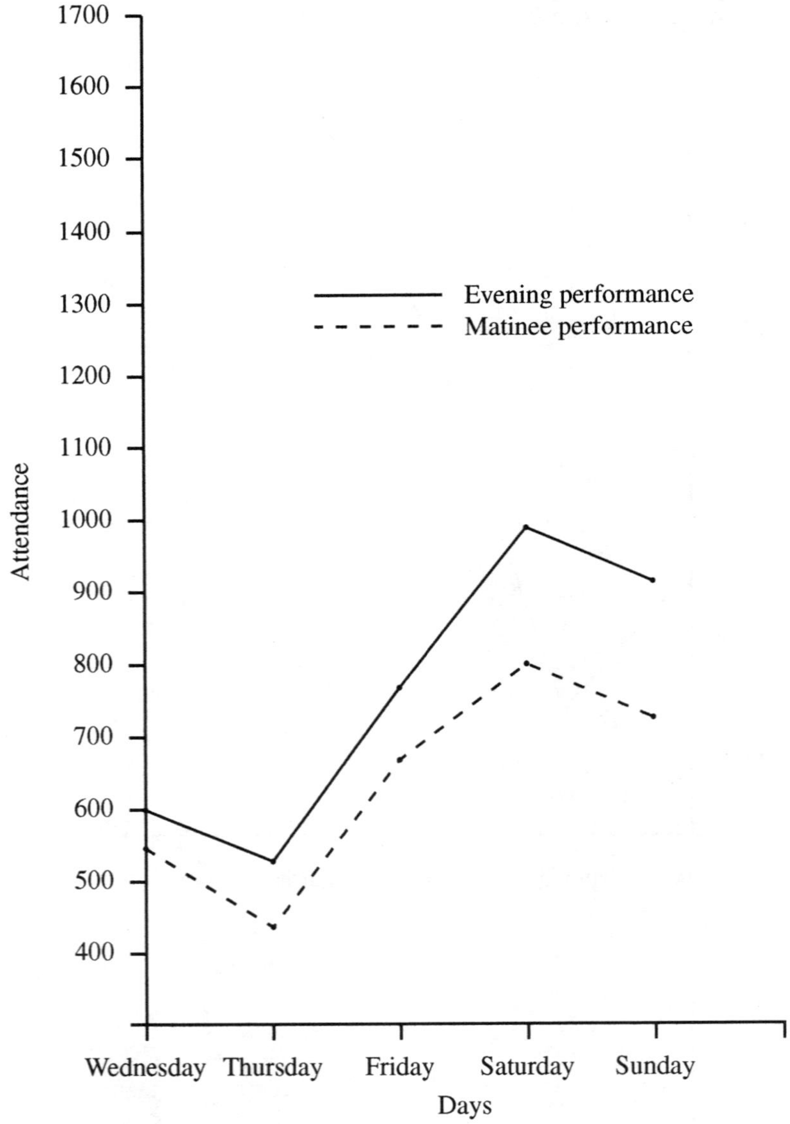
1700
1600
1500
1400
1300
1200
1100
1000
900
800
700
600
500
400
Attendance
Evening performance
Matinee performance
Wednesday
Thursday
Friday
Saturday
Sunday
Days

2. a.

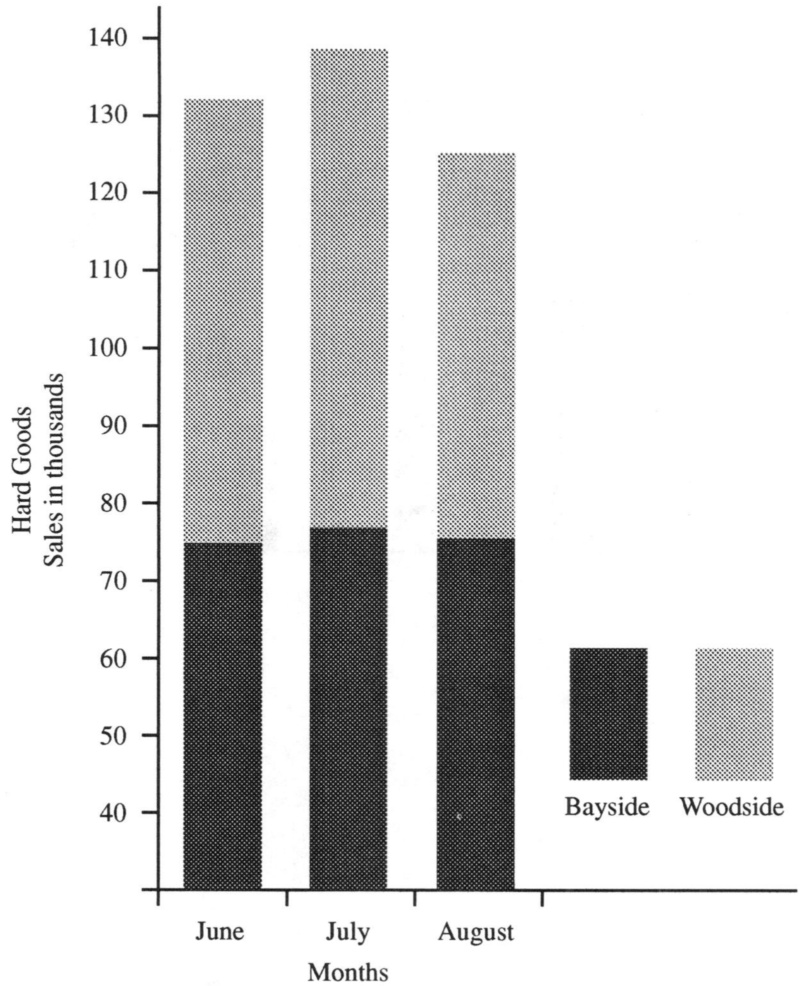
140
130
120
110
100
90
80
70
60
50
40
Hard Goods
Sales in thousands
Bayside
Woodside
June
July
August
Months

b.

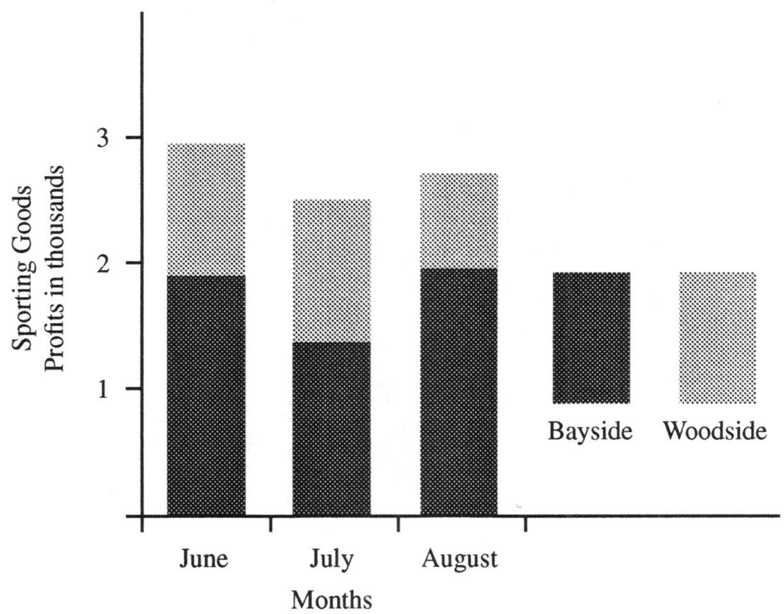
3
2
1
Sporting Goods
Profits in thousands
Bayside
Woodside
June
July
August
Months

c.

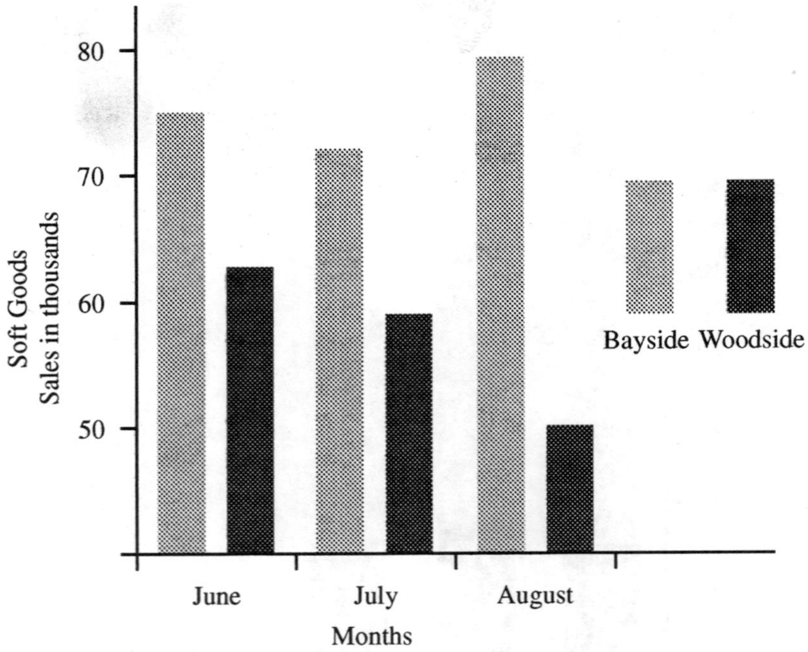
80
70
60
50
Soft Goods
Sales in thousands
June
July
August
Months
Bayside
Woodside

d.

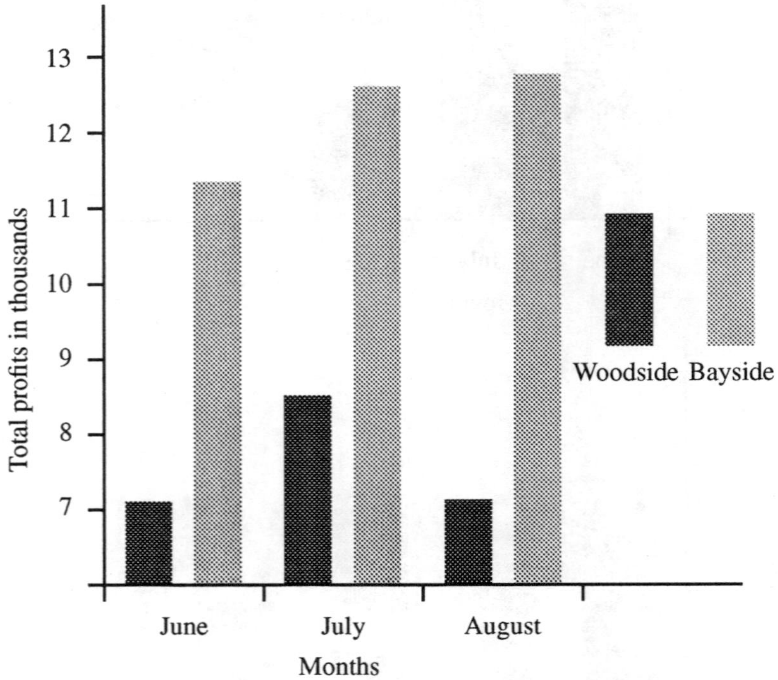
13
12
11
10
9
8
7
Total profits in thousands
June
July
August
Months
Woodside
Bayside

e. 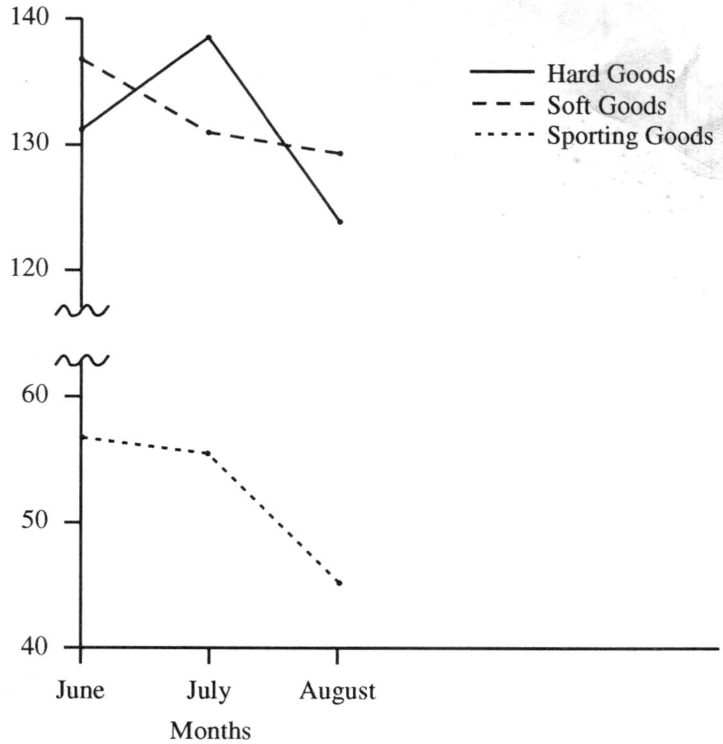
140
130
120
60
50
40
Hard Goods
Soft Goods
Sporting Goods
June
July
August
Months

3. 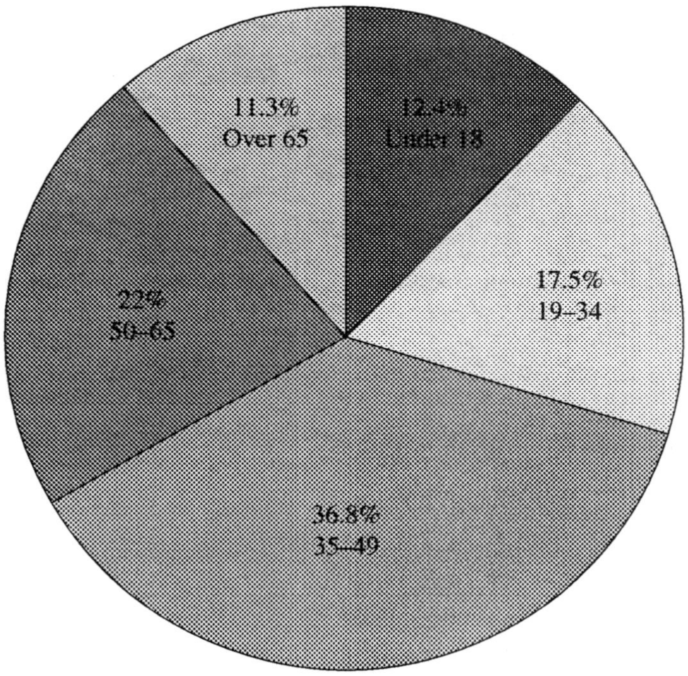
11.3%
Over 65
12.4%
Under 18
17.5%
19–34
36.8%
35–49
22%
50–65

4.

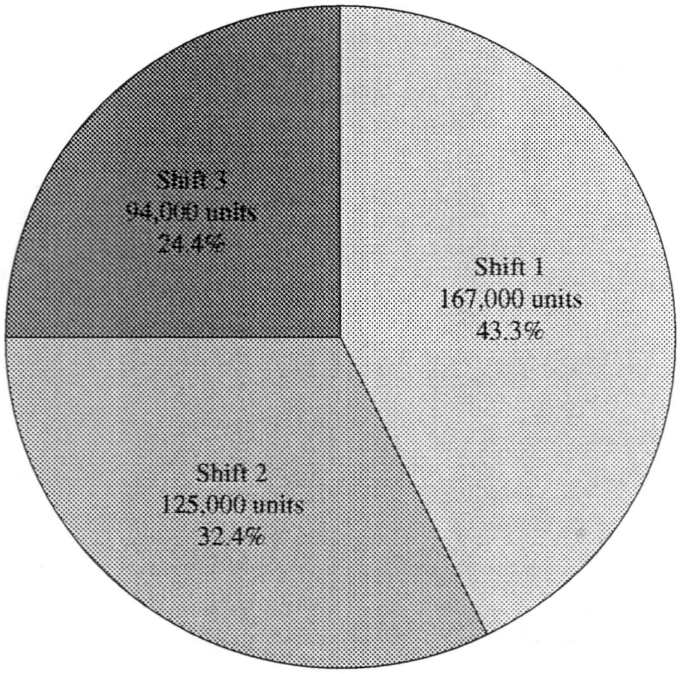

5. a. 473.2 tons
   b. 491.5 tons
   c. 490 and 493
   d. 488

6. a. 358.6 pies
   b. 330
   c. There is no mode.
   d. 403
   e. The median best describes the set, since one day's total is comparatively much higher than the others.

7. a.

Class	Frequency
51–60	1
61–70	4
71–80	17
81–90	11
91–100	9

   b. 81 mean

7. c.

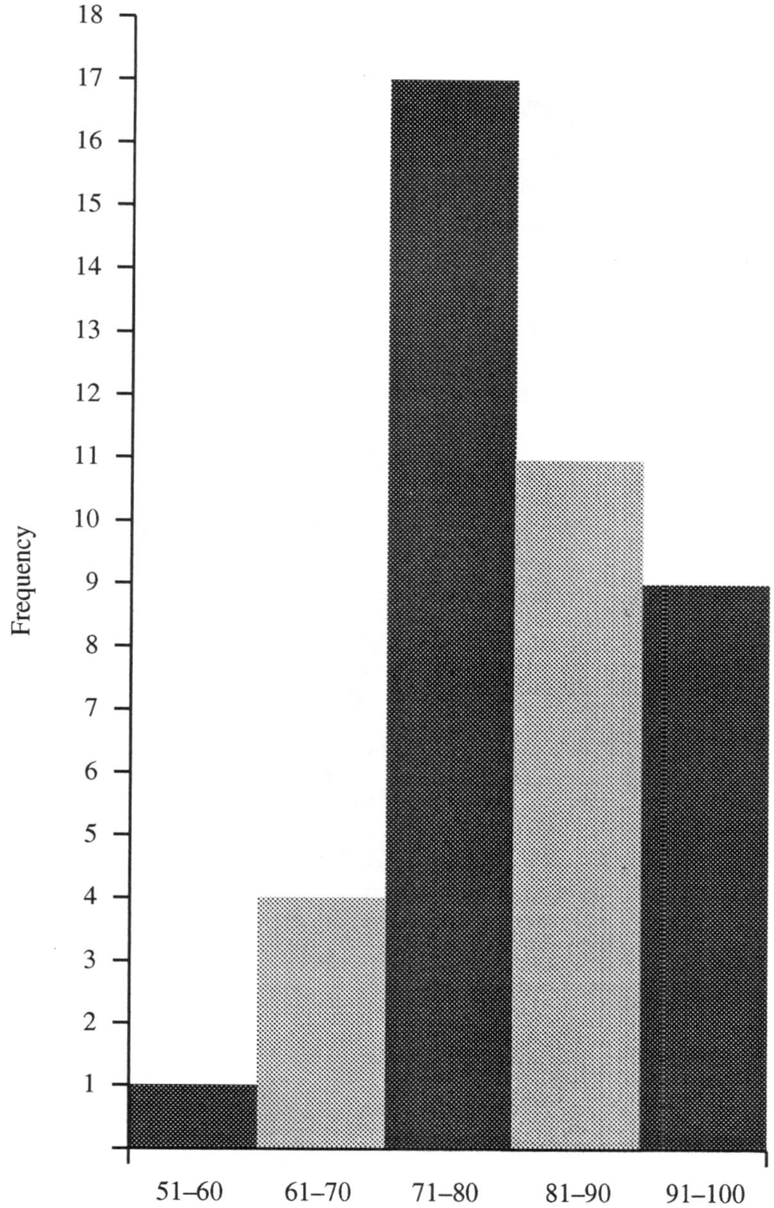

8. a.

Class	Frequency
0–999	3
1000–1999	10
2000–2999	4
3000–3999	5
4000–4999	1
5000–5999	1

b. 2,249.5 mean

c.

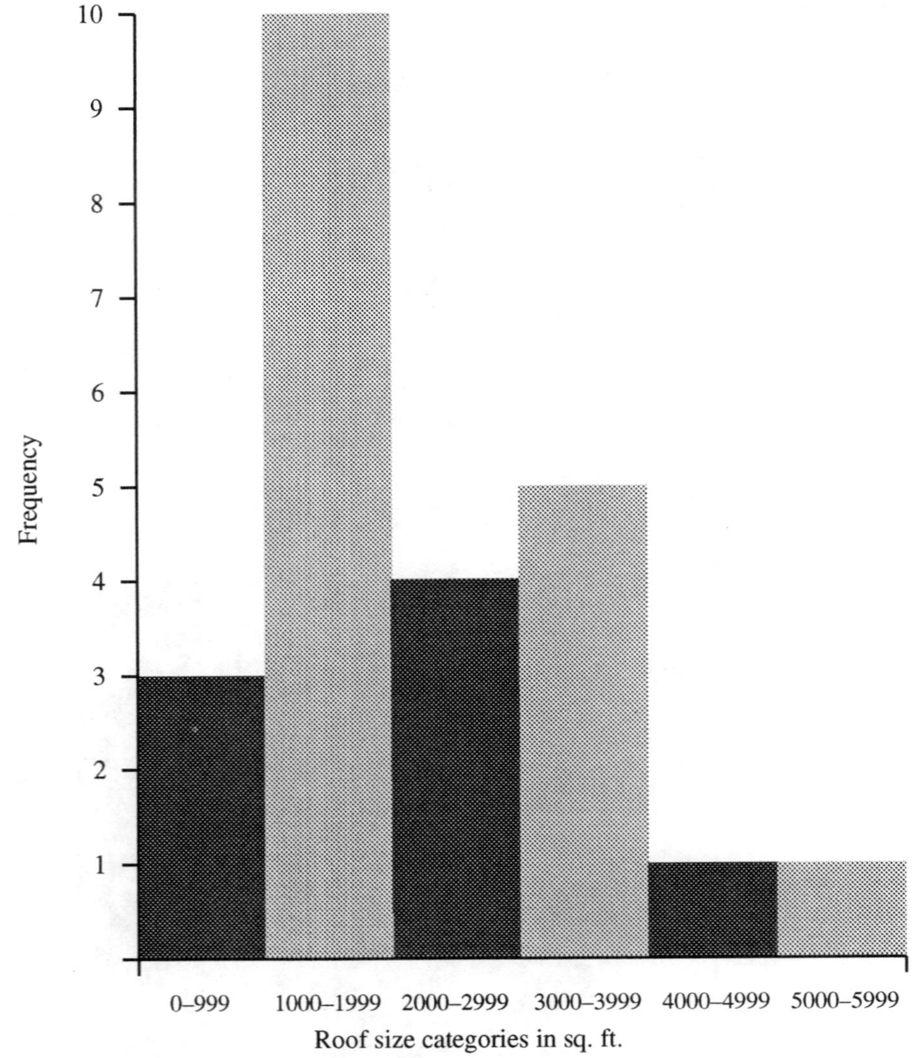
Frequency
10
9
8
7
6
5
4
3
2
1
0–999
1000–1999
2000–2999
3000–3999
4000–4999
5000–5999
Roof size categories in sq. ft.